SpringerBriefs in Mathematics

SpringerBriefs in Mathematics showcases expositions in all areas of mathematics and applied mathematics. Manuscripts presenting new results or a single new result in a classical field, new field, or an emerging topic, applications, or bridges between new results and already published works, are encouraged. The series is intended for mathematicians and applied mathematicians.

More information about this series at http://www.springer.com/series/10030

SBMAC SpringerBriefs

The **SBMAC SpringerBriefs** series publishes relevant contributions in the fields of applied and computational mathematics, mathematics, scientific computing, and related areas. Featuring compact volumes of 50 to 125 pages, the series covers a range of content from professional to academic.

The Sociedade Brasileira de Matemática Aplicada e Computacional (Brazilian Society of Computational and Applied Mathematics, SBMAC) is a professional association focused on computational and industrial applied mathematics. The society is active in furthering the development of mathematics and its applications in scientific, technological, and industrial fields. The SBMAC has helped to develop the applications of mathematics in science, technology, and industry, to encourage the development and implementation of effective methods and mathematical techniques for the benefit of science and technology, and to promote the exchange of ideas and information between the diverse areas of application.

http://www.sbmac.org.br/

José Pontes • Norberto Mangiavacchi
Gustavo R. Anjos

An Introduction to Compressible Flows with Applications

Quasi-One-Dimensional Approximation and General Formulation for Subsonic, Transonic and Supersonic Flows

 Springer

José Pontes
Faculty of Engineering
State University of Rio de Janeiro
Rio de Janeiro
Rio de Janeiro, Brazil

Norberto Mangiavacchi
Faculty of Engineering
State University of Rio de Janeiro
Rio de Janeiro
Rio de Janeiro, Brazil

Gustavo R. Anjos
COPPE/Mechanical Engineering Program
Federal University of Rio de Janeiro
Rio de Janeiro
Rio de Janeiro, Brazil

ISSN 2191-8198 ISSN 2191-8201 (electronic)
SpringerBriefs in Mathematics
ISBN 978-3-030-33252-5 ISBN 978-3-030-33253-2 (eBook)
https://doi.org/10.1007/978-3-030-33253-2

Mathematics Subject Classification: 76B07, 76G25, 76J20

This Springer imprint is published by the registered company Springer Nature Switzerland AG.
The registered company address is: Gewerbestrasse 11, 6330 Cham, Switzerland

Preface

This book is designed as an introduction to the problem of fluid flows where compressibility effects play a major role and appear not only in consequence of the primary high velocity imposed to the flow, but also in other engineering configurations, like in failures and in the opening of safety valves of pipelines employed for the transport and distribution of pressurized gases. Being devoted to applications, the authors opted to develop the material not in the standard form found in mathematical books, with results presented as proved theorems, but by deriving and simplifying the governing equations.

The first chapter presents the full three-dimensional evolution equation of potential flows, commonly found around slender aerospace structures as wings, control surface, and rockets. The chapter discusses the propagation of weak sound waves and the numerical solution of the linear wave equation in the frequency domain.

The second chapter is devoted to quasi-one-dimensional isentropic gas dynamics, and also the problem of shock compression, entropy production across shock waves, and the thickness of such waves. Still in this chapter, the book addresses the cases of flows in ducts with constant transversal section with heat addition and, in separate, with friction. The cases of flow in ducts with variable transversal section with friction or heat transfer, and with both occurring simultaneously are addressed in sequence. In these cases, the critical condition of Mach number equal to one does not occur at the minimum transversal section of the duct. The case of practical interest of isothermal flow in ducts with constant transversal section, found in pipelines is discussed next. The chapter ends pointing to the analogy between open channel hydraulics and one-dimensional gas dynamics.

Chapter 3 deals with oblique shock waves, supersonic flows over corners and wedges, and Riemman problems. Chapter 4 is devoted to the study of high velocity flows subject to small perturbations, as found in the case of slender bodies, aligned or almost aligned to the flow. The case of subsonic and supersonic flows past a periodic-shaped wall is addressed in particular. The book ends with a chapter devoted to the derivation of the basic equations of compressible fluid flows, with attention given to the one-dimensional case, where the velocity is defined as the ratio between volumetric flow and the local area of the transversal section.

At the end of each chapter, problems are proposed to further develop theoretical and computational aspects of compressible flows.

The text is suitable for advanced undergraduate students or beginning graduate students in Mathematics, Physics, and Engineering, and it may be useful for researchers in academia and in the industry.

Contents

List of Figures

Chapter 1
Compressible Potential Flows

1 Introduction

This chapter deals with compressible irrotational isentropic flows to which a velocity potential can be associated [5, 6, 10]. Viscous effects are neglected. The chapter starts with derivation of the three-dimensional, compressible, and time-dependent equation governing the evolution and the steady state potential. Particular cases of this equation are the ones governing the dynamics of weak waves, the tridimensional compressible flow close to slender bodies aligned or quasi aligned to an incident flow, Laplace's equation, which governs the incompressible potential flow under $M \ll 1$. Section 3 discusses the classification of PDEs as elliptic, parabolic, and hyperbolic.

In sequence, the chapter addresses sound propagation questions, including the numerical solution of the wave equation in one and two dimensions, in the frequency domains, using Finite Differences and Finite Elements methods.

2 The Equation of Compressible Potential Flows

Irrotational flows admit a velocity potential ϕ, from which the velocity field **v** derives:

$$\mathbf{v} = \mathbf{grad}\,\phi. \tag{1.1}$$

In addition, irrotational flows satisfy Bernoulli's equation, given by Eq. 5.11. Conditions for an irrotational flow to stay in this condition are given by Kelvin's theorem Eq. 5.14 which states that vorticity is produced by thermodynamic and viscous effects. We assume that these conditions are satisfied in the derivation of

© The Author(s), under exclusive licence to Springer Nature Switzerland AG 2019
J. Pontes et al., *An Introduction to Compressible Flows with Applications*,
SpringerBriefs in Mathematics, https://doi.org/10.1007/978-3-030-33253-2_1

the evolution equation of the time-dependent potential associated with compressible irrotational flows. We derive now an equation for the time evolution of $\phi = \phi(\mathbf{x}, t)$.

Euler's equation can be written as:

$$\frac{\partial \mathbf{v}}{\partial t} + \mathbf{grad}\, \frac{v^2}{2} - \mathbf{v} \times \mathbf{rot}\, \mathbf{v} = -\frac{1}{\rho} \mathbf{grad}\, p,$$

where $v^2 = \mathbf{v} \cdot \mathbf{v}$. In the case of irrotational flows, this equation becomes:

$$\frac{\partial v_i}{\partial t} + \frac{\partial}{\partial x_i} \frac{v^2}{2} = -\frac{1}{\rho} \frac{\partial p}{\partial x_i}.$$

Expressing the velocity components as functions of the velocity potential, $v_i = \partial \phi / \partial x_i$:

$$\frac{\partial}{\partial t} \frac{\partial \phi}{\partial x_i} + \frac{1}{\rho} \frac{\partial p}{\partial x_i} + \frac{\partial}{\partial x_i} \frac{v^2}{2} = 0.$$

Upon integrating this last equation, we have:

$$\int \frac{\partial}{\partial t} \frac{\partial \phi}{\partial x_i} dx_i + \int \frac{1}{\rho} \frac{\partial p}{\partial x_i} dx_i + \int \frac{\partial}{\partial x_i} \frac{v^2}{2} dx_i = \frac{\partial \phi}{\partial t} + \int \frac{dp}{\rho} + \frac{v^2}{2} = F(t).$$

The time dependency expressed by the function $F(t)$ can be incorporated to the potential ϕ, leading to:

$$\frac{\partial \phi}{\partial t} + \int \frac{dp}{\rho} + \frac{v^2}{2} = 0.$$

We derive this last equation with respect to time, to obtain:

$$\frac{\partial^2 \phi}{\partial t^2} + \frac{\partial}{\partial t} \int \frac{dp}{\rho} + \frac{1}{2} \frac{\partial v^2}{\partial t} = 0. \qquad (1.2)$$

At this point, we introduce the hypothesis of isentropic flow, by further assuming that no heat is transferred between the fluid particles, and that no heat is generated in the medium. In consequence of these assumptions, along with the hypothesis of inviscid flow, the equation of the entropy, given by:

$$T \frac{Ds}{Dt} = \frac{1}{\rho} \tau : \mathbf{grad}\, \mathbf{v} + \frac{\kappa}{\rho} \nabla^2 T + \frac{\dot{Q}}{\rho},$$

simplifies and takes the form:

$$T \frac{Ds}{Dt} = 0,$$

i.e., the flow is isentropic. At each point of the field, we have, thus:

$$p = p(\rho, s)$$

$$dp = \left(\frac{\partial p}{\partial \rho}\right)_s d\rho + \left(\frac{\partial p}{\partial s}\right)_\rho ds = \left(\frac{\partial p}{\partial \rho}\right)_s d\rho = a^2 d\rho, \qquad (1.3)$$

where $a^2 = (\partial p/\partial \rho)_s$ and $ds = 0$. The term containing the integral in Eq. 1.2 can be rewritten as:

$$\frac{\partial}{\partial t} \int \frac{dp}{\rho} = \frac{\partial}{\partial t} \int \frac{a^2}{\rho} d\rho = a^2 \frac{\partial}{\partial t} \int \frac{d\rho}{\rho} = a^2 \frac{\partial}{\partial t} \ln \rho = \frac{a^2}{\rho} \frac{\partial \rho}{\partial t}.$$

The term $a^2 = (\partial p/\partial \rho)_s$ has units of square of a velocity. In fact,

$$\left[\left(\frac{\partial p}{\partial \rho}\right)_s\right] = \left[\frac{F}{L^2 \times \rho}\right] = \left[\frac{M \times L}{t^2}\right]\left[\frac{1}{L^2}\right]\left[\frac{L^3}{M}\right] = \left[\frac{L^2}{t^2}\right].$$

In the case of perfect gases, we have:

$$a^2 = \left(\frac{\partial p}{\partial \rho}\right)_s = \gamma RT.$$

As we shall see below, a^2 is the square of the sound velocity. By introducing this definition in Eq. 1.2 we have:

$$\frac{\partial^2 \phi}{\partial t^2} + \frac{a^2}{\rho} \frac{\partial \rho}{\partial t} + \frac{1}{2} \frac{\partial v^2}{\partial t} = 0. \qquad (1.4)$$

Upon multiplying Euler's equation by v_i and mentioning that $v_i \partial v_i / \partial t = 1/2 \, \partial v^2 / \partial t$, we obtain:

$$\frac{1}{2} \frac{\partial v^2}{\partial t} + v_i v_j \frac{\partial v_i}{\partial x_j} = -\frac{v_i}{\rho} \frac{\partial p}{\partial x_i}$$

or

$$\frac{1}{2} \frac{\partial v^2}{\partial t} + v_i v_j \frac{\partial v_i}{\partial x_j} = -\frac{a^2}{\rho} v_i \frac{\partial \rho}{\partial x_i},$$

since $\partial p / \partial x_i = a^2 \partial \rho / \partial x_i$. By using the continuity equation, we replace the term $v_i \partial \rho / \partial x_i$ in the above equation by:

$$v_i \frac{\partial \rho}{\partial x_i} = -\left(\frac{\partial \rho}{\partial t} + \rho \frac{\partial v_i}{\partial x_i}\right)$$

to obtain:

$$\frac{1}{2}\frac{\partial v^2}{\partial t} + v_i v_j \frac{\partial v_i}{\partial x_j} = \frac{a^2}{\rho}\left(\frac{\partial \rho}{\partial t} + \rho \frac{\partial v_i}{\partial x_i}\right).$$

Taking Eq. 1.4 into account, we replace the term $(a^2/\rho)(\partial\rho/\partial t)$ in the above equation:

$$\frac{1}{2}\frac{\partial v^2}{\partial t} + v_i v_j \frac{\partial v_i}{\partial x_j} = a^2 \frac{\partial v_i}{\partial x_i} - \frac{\partial^2 \phi}{\partial t^2} - \frac{1}{2}\frac{\partial v^2}{\partial t}.$$

By replacing $1/2\, \partial v^2/\partial t = v_i \partial v_i/\partial t$ we have:

$$2v_i\frac{\partial v_i}{\partial t} + v_i v_j \frac{\partial v_i}{\partial x_j} = a^2 \frac{\partial v_i}{\partial x_i} - \frac{\partial^2 \phi}{\partial t^2}.$$

We replace now $v_i = \partial\phi/\partial x_i$ to obtain:

$$2\frac{\partial \phi}{\partial x_i}\frac{\partial^2 \phi}{\partial t \partial x_i} + \frac{\partial \phi}{\partial x_i}\frac{\partial \phi}{\partial x_j}\frac{\partial^2 \phi}{\partial x_i \partial x_j} = a^2 \frac{\partial^2 \phi}{\partial x_i^2} - \frac{\partial^2 \phi}{\partial t^2},$$

and finally:

$$\frac{\partial^2 \phi}{\partial x_i^2} = \frac{1}{a^2}\left(\frac{\partial \phi}{\partial x_i}\frac{\partial \phi}{\partial x_j}\frac{\partial^2 \phi}{\partial x_i \partial x_j} + 2\frac{\partial \phi}{\partial x_i}\frac{\partial^2 \phi}{\partial x_i \partial t} + \frac{\partial^2 \phi}{\partial t^2}\right). \tag{1.5}$$

In vector notation:

$$\nabla^2\phi = \frac{1}{a^2}\left(\nabla\phi \otimes \nabla\phi : \nabla\nabla\phi + 2\nabla\phi \cdot \frac{\partial \nabla\phi}{\partial t} + \frac{\partial^2 \phi}{\partial t^2}\right). \tag{1.6}$$

The potential equation of the compressible flows is written, in indexes and in vector notations, by Eqs. 1.5 and 1.6 respectively. The equation holds in domains without entropy production and where the inflow is irrotational.

Boundary conditions for the potential ϕ:

1. Stationary solids:

$$\frac{\partial \phi}{\partial n} = 0,$$

where n is the coordinate along the direction $\mathbf{n}$, perpendicular to the solid surface. The condition is equivalent to $\mathbf{grad}\,\phi \cdot \mathbf{n} = 0$.

2. Solids moving with steady velocity $\mathbf{U}$:

$$\frac{\partial \phi}{\partial n} = \mathbf{n} \cdot \mathbf{U}.$$

This condition may also be written as: $\mathbf{n} \cdot (\mathbf{v} - \mathbf{U}) = 0$, where $\mathbf{v}$ is local velocity.

We consider now some limit cases for Eq. 1.5.

Notation We denote the derivation by an index containing the variable relative to which the potential ϕ is derived: $\partial \phi \partial t = \phi_t$, $\partial^2 \phi / \partial x^2 = \phi_{xx}$, etc.

1. Two-dimensional steady flows:

$$(\phi_x^2 - a^2)\phi_{xx} + (\phi_y^2 - a^2)\phi_{yy} + 2\phi_x\phi_y\phi_{xy} = 0 \tag{1.7}$$

or

$$(u^2 - a^2)\phi_{xx} + (v^2 - a^2)\phi_{yy} + 2v_x v_y \phi_{xy} = 0,$$

where u and v stand for the velocity components along the x and y directions, respectively.
2. Tridimensional steady flows, with $v_x \gg v_y$, $v_x \gg v_z$ (slender bodies with low angle of attack and out of the transonic region):

$$(1 - M^2)\phi_{xx} + \phi_{yy} + \phi_{zz} = 0.$$

In this case, the slender body aligned or quasi aligned to the inflow gives raise to small velocity components in the orthogonal directions.
3. Steady tridimensional flow, with all velocity components small, when compared to the sound velocity. In this case we consider $1/a^2 \longrightarrow 0$ to obtain:

$$\phi_{xx} + \phi_{yy} + \phi_{zz} = 0.$$

The last one is the well-known Laplace's equation, which can also be derived from the continuity equation for incompressible flows, also assuming that irrotational flows admit a potential from which the velocity field derives (Eq. 1.1). By replacing this equation in the continuity equation we have:

$$\frac{\partial v_i}{\partial x_i} = \frac{\partial}{\partial x_i}\frac{\partial \phi}{\partial x_i} = \nabla^2 \phi = 0. \tag{1.8}$$

We stress that the hypothesis of incompressible flows is equivalent to assume that the sound velocity diverges. The right-hand side expression of Eq. 1.5 vanishes. Equation 1.8 applies to any incompressible potential flow and not only to cases where $v_x \gg v_y$ and $v_x \gg v_z$. Problems to which this equation applies belong to a particular class where the hydrodynamic field is obtained by solving the potential equation, followed by a post-processing where the velocity components are obtained as the gradient of the potential. When necessary, the pressure gradient is obtained with the Euler equation.

Equation 1.8 has the important feature of being linear. Due to that, the sum of two solutions of the potential equation is also a solution of Laplace's equation. The superposition principle applies.

4. Unsteady one-dimensional flows

$$\phi_{xx} = \frac{1}{a^2} \left(\phi_x^2 \phi_{xx} + \phi_{tt} + 2\phi_x \phi_{tx} \right)$$

or

$$\phi_{xx} = \frac{1}{a^2} \left(v_x^2 \phi_{xx} + \phi_{tt} + 2v_x \phi_{tx} \right).$$

5. Unsteady three-dimensional flows, $v_i \ll a^2$:

$$\nabla^2 \phi = \frac{1}{a^2} \phi_{tt}. \tag{1.9}$$

We have in this case the weak waves equation, which governs the evolution of small perturbations propagating in the medium with velocity a, namely, a is the sound velocity (see also Sect. 4).

3   A Classification of Partial Differential Equations

We consider a steady two-dimensional potential flow and assume that ϕ and **grad** ϕ are known on a given curve Σ. We rewrite Eq. 1.7 as:

$$A\phi_{xx} + B\phi_{xy} + C\phi_{yy} + D = 0,$$

with independent variables x and y, dependent variable ϕ, and coefficients A, B, C, and D, as functions of a, x, y, ϕ, ϕ_x, and ϕ_y. Let us assume that $x = x(\sigma)$ and $y = y(\sigma)$. We denote:

$$p = \phi_x \qquad r = \phi_{xx}$$
$$q = \phi_y \qquad s = \phi_{xy} \qquad t = \phi_{yy}.$$

We have, then:

$$Ar + Bs + Ct = -D$$
$$\frac{dp}{d\sigma} = \frac{d\phi_x}{d\sigma} = \frac{dx}{d\sigma}r + \frac{dy}{d\sigma}s$$

$$\frac{dq}{d\sigma} = \frac{d\phi_y}{d\sigma} = \frac{dx}{d\sigma}s + \frac{dy}{d\sigma}t.$$

By rewriting the above system of equations in matrix form:

$$\begin{pmatrix} A & B & C \\ \dfrac{dx}{d\sigma} & \dfrac{dy}{d\sigma} & 0 \\ 0 & \dfrac{dx}{d\sigma} & \dfrac{dy}{d\sigma} \end{pmatrix} \begin{pmatrix} r \\ s \\ t \end{pmatrix} = \begin{pmatrix} -D \\ \dfrac{dp}{d\sigma} \\ \dfrac{dq}{d\sigma} \end{pmatrix}.$$

Let us find conditions for the discontinuity of the velocity derivatives and, in consequence the conditions for the divergence of higher order derivatives, ϕ_{xx}, ϕ_{xy}, and ϕ_{yy}. Let σ be the parameter of a characteristic curve, along which the highest order derivatives namely the velocity derivatives are discontinuous. We prescribe this condition by imposing that the determinant of the matrix of coefficients of the above equation vanishes. This condition writes as:

$$A\left(\frac{dy}{d\sigma}\right)^2 - B\left(\frac{dx}{d\sigma}\right)\left(\frac{dy}{d\sigma}\right) + C\left(\frac{dx}{d\sigma}\right)^2 = 0$$

or, dividing the last equation by $(dx/d\sigma)^2$:

$$A\left(\frac{dy}{dx}\right)^2 - B\left(\frac{dy}{dx}\right) + C = 0,$$

from where we have:

$$\frac{dy}{dx} = \frac{B \pm \sqrt{B^2 - 4AC}}{2A}.$$

The derivatives of the velocity field are continuous when the roots of the equation are imaginaries, and discontinuous if real. Equation 1.7 may be classified in three groups, depending on the value of the associated parameters:

1. Parabolic, if $B^2 - 4AC = 0$;
2. Hyperbolic, if $B^2 - 4AC > 0$;
3. Elliptic, if $B^2 - 4AC < 0$ (no characteristic curves exist in the field, along which the derivatives of the velocity are discontinuous);

Some examples:

1. Laplace's equation:

$$\frac{\partial^2 \phi}{\partial x^2} + \frac{\partial^2 \phi}{\partial y^2} = 0.$$

No real characteristics exist in the fields obeying Laplace's equation. All the derivatives are continuous.

2. One-dimensional, time-dependent temperature equation

$$\frac{\partial T}{\partial t} = \frac{\partial^2 T}{\partial x^2}.$$

This equation presents one family of characteristics. The equation is parabolic.
3. Wave equation:

$$\frac{\partial^2 \phi}{\partial t^2} = a^2 \frac{\partial^2 \phi}{\partial x^2}.$$

This equation presents two families of characteristics, along the straight lines $dx/dt = \pm a$. The field always presents a a discontinuous region. The equation is, in this case, hyperbolic.
4. In the case of Eq. 1.7, written in the form:

$$\left(u^2 - a^2\right)\phi_{xx} + 2uv\phi_{xy} + \left(v^2 - a^2\right)\phi_{yy} = 0,$$

where $u = \phi_x$, $v = \phi_y$, $A = \left(u^2 - a^2\right)$, $B = 2uv$, and $C = \left(v^2 - a^2\right)$,

we have:

$$B^2 - 4AC = a^2\left[\left(u^2 + v^2\right) - a^2\right].$$

If $|\mathbf{v}|^2 = \left(u^2 + v^2\right) < a^2$, i.e., if the flow regime is subsonic, the equation is elliptic and there are no discontinuities at the velocity derivatives. If $|\mathbf{v}|^2 = \left(u^2 + v^2\right) > a^2$, namely, if the flow is supersonic, $B2 - 4AC > 0$, the equation is hyperbolic. In this case:

$$\frac{dy}{dx} = \frac{uv \pm \sqrt{a^2\left(u^2 + v^2 - a^2\right)}}{u^2 - a^2} = \frac{\dfrac{v}{u} \pm \dfrac{a^2}{u^2}\sqrt{M^2 - 1}}{1 - \dfrac{a^2}{u^2}} = \tan\left(\alpha \pm \theta\right),$$

$$(1.10)$$

where:

$$M = \frac{\sqrt{u^2 + v^2}}{a} \qquad \tan\alpha = \frac{v}{u} \qquad \sin\theta = \frac{a}{\sqrt{u^2 + v^2}} = \frac{1}{M}.$$

If $\mathbf{v} = 0$, Eq. 1.10 takes the following simplified form:

$$\frac{dy}{dx} = \tan\left(\pm\theta\right) = \pm\frac{1}{\sqrt{M^2 - 1}},$$

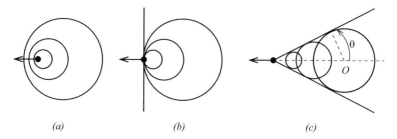

Fig. 1.1 Characteristic lines of a steady two-dimensional compressible flow. (**a**): A point moving with subsonic velocity emits small perturbations that propagate faster than the point; no discontinuities exist in the velocity derivatives. The equation of the potential is elliptic. (**b**): The point moves with the sound velocity. A family of characteristics exists, along which the velocity derivatives are discontinuous. The region ahead (at left) of the characteristics is the silent zone and does not receive signals emitted by the point. The equation of the potential is parabolic. (**c**): The point moves with supersonic velocity. The field admits two families of characteristics, along which the velocity derivatives are discontinuous. Points located ahead of the characteristics are in the zone of silence. The equation of the potential is hyperbolic. A small perturbation initially at the point O stays in the characteristic line defining the Mach's Cone and moves perpendicularly to it, whereas the particle emitting perturbations move attached to the cone's vertex

where θ is the angle between the characteristics and the velocity vector of a point moving in the field. We also have:

$$\sin\theta \; = \; \frac{a}{u} \; = \; \frac{1}{M}, \quad \text{and} \quad \cos\theta \; = \; \frac{\sqrt{M^2 - 1}}{M}.$$

Figure 1.1c shows that, at the end of an elapsed time equal to one, a perturbation emitted by a point moving with velocity u covers a distance numerically equal to a, while the point covers a distance numerically equal to u. From the same figure we infer that $\sin\theta = a/u = 1/M$.

For a point moving with $v = u$, the characteristic lines along which the velocity derivatives are discontinuous and the regions of the field receiving signals of small perturbations emitted by the point are shown schematically in Fig. 1.1. The inner region delimited by the characteristics is named as the *Mach's Cone*.

4 Weak Waves: Sound Velocity

In the previous section we showed that the potential equation of compressible flows (Eq. 1.5) reduces, in the case of the evolution of weak perturbations of rest states, to the wave equation for the flow potential ϕ (Eq. 1.9). Weak perturbations of the rest state in a gas are, by definition, sound waves. In addition, for a perfect gas, we have:

$$a^2 \; = \; \gamma RT.$$

We show now that not only perturbations of the rest potential, but also, of velocity, specific mass and pressure perturbations do obey the equation of weak waves and propagate with velocity a, namely, a is effectively the sound velocity.

The potential ϕ can be written in the form of a sum of the uniform, time-independent one, $\bar{\phi}$, plus a perturbation, $\tilde{\phi}$:

$$\phi = \bar{\phi} + \tilde{\phi},$$

with the uniform and time-independent $\bar{\phi}$, trivially obeying Eq. 1.9. The perturbation term $\tilde{\phi}$ also satisfies the same equation:

$$\nabla^2 \tilde{\phi} = \frac{1}{a^2} \tilde{\phi}_{tt}.$$

Upon applying the gradient operator to this last equation we find:

$$\nabla^2 \frac{\partial \tilde{\phi}}{\partial x_i} = \frac{1}{a^2} \frac{\partial^2}{\partial t^2} \frac{\partial \tilde{\phi}}{\partial x_i},$$

or

$$\nabla^2 \tilde{v}_i = \frac{1}{a^2} \frac{\partial^2 \tilde{v}_i}{\partial t^2}, \tag{1.11}$$

where $\tilde{v}_i$ is the velocity perturbation, associated with $\tilde{\phi}$. In order to prove that specific mass perturbations around a steady value $\bar{\rho}$, we consider the three-dimensional Euler's equation applicable to a rest fluid submitted to a small perturbation. Specific mass and pressure are written as:

$$\rho = \bar{\rho} + \tilde{\rho}$$

$$p = \bar{p} + \tilde{p},$$

where $\bar{p}$ and $\bar{\rho}$ are the steady unperturbed values of pressure and specific mass, whereas $\tilde{p}$ and $\tilde{\rho}$ stand for the time-dependent perturbations of the steady state variables. Assuming isentropic flow and neglecting gravitational effects, we write Euler's equation in the form:

$$(\bar{\rho} + \tilde{\rho}) \left[\frac{\partial}{\partial t} (\bar{v}_i + \tilde{v}_j) + (\bar{v}_j + \tilde{v}_j) \frac{\partial}{\partial x_j} (\bar{v}_i + \tilde{v}_i) \right] = -a^2 \frac{\partial \rho}{\partial x_i}.$$

Euler's equations are now linearized. Reminding that the base state variables are uniform in space and time independent, that perturbations are small, and, additionally, that $\bar{v}_i \equiv 0$, we obtain the linearized isentropic Euler's equations:

$$\bar{\rho} \frac{\partial \tilde{v}_i}{\partial t} + a^2 \frac{\partial \tilde{\rho}}{\partial x_i} = 0. \tag{1.12}$$

By deriving Eq. 1.12 with respect to time and rearranging terms, and using Eq. 1.11 we obtain:

$$\nabla^2 \tilde{v}_i = \frac{1}{a^2} \frac{\partial^2 \tilde{v}_i}{\partial t^2} = -\frac{1}{\bar{\rho}} \frac{\partial^2 \tilde{\rho}}{\partial t \, \partial x_i}.$$

Derivation of the above equation with respect to time leads to:

$$\nabla^2 \frac{\partial \tilde{v}_i}{\partial t} = -\frac{1}{\bar{\rho}} \frac{\partial^3 \tilde{\rho}}{\partial t^2 \, \partial x_i}. \tag{1.13}$$

From Eq. 1.12 we obtain:

$$\frac{\partial \tilde{v}_i}{\partial t} = -\frac{a^2}{\bar{\rho}} \frac{\partial \tilde{\rho}}{\partial x_i}. \tag{1.14}$$

Upon replacing the expression of $\partial v_i / \partial t$, given by Eq. 1.14 into Eq. 1.13 we find:

$$\nabla^2 \frac{\partial \rho}{\partial x_i} = \frac{1}{a^2} \frac{\partial^3 \tilde{\rho}}{\partial t^2 \, \partial x_i}. \tag{1.15}$$

Integration with respect to x_i leads to the sought wave equation for the specific mass perturbation:

$$\nabla^2 \rho = \frac{1}{a^2} \frac{\partial^2 \tilde{\rho}}{\partial t^2}. \tag{1.16}$$

The pressure perturbation $\tilde{p}$ also obeys a wave equation. In fact, for an isentropic process we can write:

$$d\tilde{\rho} = \frac{1}{a^2} d\tilde{p}.$$

Replacing the above expression of $d\tilde{\rho}$ in Eq. 1.16 leads to the wave equation for the small pressure perturbations:

$$\nabla^2 \tilde{p} = \frac{1}{a^2} \frac{\partial^2 \tilde{p}}{\partial t^2}. \tag{1.17}$$

5 Solution of the Wave Equation in the Frequency Domain

In this section we describe the method for finding time periodic solutions of the pressure perturbation equation. As a first step we assume that the solution of Eq. 1.17 may be written as a superposition of harmonic modes in the form:

$$p = \hat{p} \, e^{j\omega t},$$

where $\hat{p}$ stems for the pressure scalar field in the frequency domain, p, for the pressure field in the time domain. At the frequency domain $j = \sqrt{-1}$. In order to analyze the system response to periodic forcing, we consider the solutions associated with a given frequency ω. Upon replacing the above solution in Eq. 1.16 and dividing the result by $e^{j\omega t}$ we obtain a Helmholtz equation identical to the above referred with time derivatives replaced by $j\omega$:

$$\kappa^2 p + \nabla^2 p = 0, \qquad (1.18)$$

where $\kappa^2 = \frac{1}{a_0^2}\omega^2$ is the square of the wave number, and a_0 the sound velocity. In the case of acoustic waves propagating in the air at $0\,°C$, at the sea level, typical values of the parameters are $\gamma = 1.4$, $\rho_0 = 1.29\,\text{kg/m}^3$, and the $P_0 = 1.01 \times 10^5\,\text{n/m}^2$ for the atmospheric pressure, leading to $a_0 = 330\,\text{m/s}$.

Formulation and solution of the eigenvalue problem in the frequency domain in one and two dimensions using the Finite Differences and Finite Elements Methods, and using both the strong and the weak formulations are presented in the exercises at the end of this chapter.

6 Problems

1. **Formulation of the eigenvalue problem in the frequency domain, in one dimension, using the strong form:**
   Show that the one-dimensional pressure wave equation writes in the form:

$$\kappa^2 \hat{p} + D^2 \hat{p} = 0,$$

where $D = d/dx$. Consider now the problem of finding if the natural frequencies of a one-dimensional rectangular cavity, open at one end and close at the other, is close to a reference frequency $\omega_0 = 2\pi f_0$. The problem also involves the identification of the vibration modes associated with the natural frequencies. Define the nondimensional wave numbers and natural frequencies, by $\kappa^* = 2\pi f_0^*$ and $f_0^* = \frac{f}{f_0}$, drop the asterisks, and rewrite the eigenvalue-eigenfunction problem as:

$$D^2 \hat{p} = -\omega^2 \hat{p}. \qquad (1.19)$$

The eigenvalues $\omega = \kappa/a_0^2$ represent the resonant angular frequencies of the cavity, associated with the steady waves, of the transient solution. Boundary conditions at a rigid surface with a local normal vector $\mathbf{n}$ require

$$\mathbf{n} \cdot \nabla p = 0. \qquad (1.20)$$

Show that in the one-dimensional case, occurring, for instance, when we consider the sound propagation in a tube closed at one end, boundary conditions reduce to:

$$\frac{\partial p}{\partial x}\Big|_{x=0} = 0 \quad \text{(for the closed end at } x = 0),$$

$$p(L) = 0 \quad \text{(for the open end at } x = L).$$

2. **Discrete formulation and resolution of the one-dimensional eigenvalue-eigenfunction problem in Finite Differences:**
Finite Differences discretization of the one-dimensional real eigenvalue problem replaces the space derivatives by a discrete representation. If a centered second order differencing scheme is adopted and node points are uniformly distributed, $x_i = x_0 + \Delta x\, i$, $i = 0, 1, 2, \ldots, n_x$, the discrete formula reads:

$$\frac{d^2 p}{dx^2} = \frac{p_{i-1} - 2p_i + p_{i+i}}{\Delta x^2} + \mathcal{O}(\Delta x^2). \tag{1.21}$$

Verify that upon replacing the space derivative by Eq. 1.21, one obtains a linear system of equations, written in matrix form, as:

$$A p^h = -\kappa^2 M p^h,$$

where nonvanishing elements of the matrices A and M read:

$$A(i, i-1) = \frac{1}{\Delta x^2}, \quad A(i, i) = -\frac{2}{\Delta x^2}, \quad A(i, i+1) = \frac{1}{\Delta x^2}, \quad M(i, i) = -1.$$

A and M are the rigidity and M mass matrices, respectively, and p^h stems for the pressure at the computational nodes.

Define a computational mesh, including information about the boundary conditions and the medium properties, and solve the eigenvalue-eigenfunction problem associated with Eq. 1.19, using the following algorithm:

```
clear;clc;
% Pre-processing: Domain Definition
nnodes=100; % number of nodes
Xmin=0; Xmax=1;
dx=(Xmax-Xmin)/(nnodes-1);
x=Xmin:dx:Xmax;

% Assembling matrices (operator d^2/dx^2)
for i=2:nnodes-1
    A(i,i-1)=1/(dx*dx);
    A(i,i)=-2/(dx*dx);
    A(i,i+1)=1/(dx*dx);
    M(i,i)=-1;
end
```

```
% Application of Dirichlet BCs
A(1,1)=1;
M(1,1)=0;
A(nnodes,nnodes)=1;
M(nnodes,nnodes)=0;

% Application of Neumann BCs
A(nnodes,nnodes)=1;
A(nnodes,nnodes-1)=-1;

% Solving eigenvalues problem
[V,D] = eigs(A,M,6,'sm');

% Plotting first 4 eigenfunctions
plot(x,V(:,1)',x,V(:,2)',x,V(:,3)',x,V(:,4)');
xlabel('domain lenght x','FontSize', 20)
xtickformat('%.1f')
ylabel('amplitude','FontSize', 20)
ytickformat('%.2f')
set(gca,'FontSize',20)
```

Validation Tests: Resonance in a Tube

Validate the methodology by evaluating the eigenvalues and eigenfunctions associated with a steady wave in a tube, in order to confirm the exactitude of the method. A possible algorithm written in *Octave* language, for defining the computational mesh, assembling the matrices, with the boundary conditions, and solving the problem is given below. Validate the above methodology by evaluating the eigenvalues and eigenfunctions associated with a steady wave in a tube, and show that modes present the form shown in Fig. 1.2.

Fig. 1.2 Eigenfunctions associated with a steady one-dimensional acoustic wave in a tube open at the left end and closed at $x = L$. The results show that the eigenfunctions are Fourier modes satisfying the Dirichlet BC at left and the Neumann ones at right

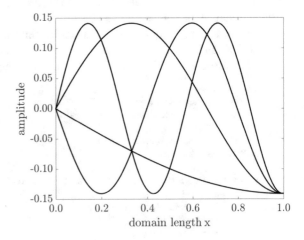

3. **Discrete formulation of the eigenvalue-eigenfunction problem in one dimension with the Finite Elements Method:**

Galerkin formulation of eigenvalue-eigenfunction problems consists in choosing base and weighing functions in finite dimensional spaces, $\mathscr{P}^h$ and $\mathscr{W}^h$. The discrete problem turns in finding $p^h \in \mathscr{P}^h$ and $\omega \in \mathscr{R}$ such that for all $w^h \in \mathscr{W}^h$:

$$\left(Dp^h, Dw^h\right) = \left(\omega p^h, \omega w^h\right).$$

By using Finite Element canonical basis in Cartesian coordinates, show that the following linear system, written in matrix form, is obtained:

$$AP^h = -\omega^2 M P^h,$$

where:

$$A_{ij} = -\left(Dw_i^h, Dw_j^h\right) \quad \text{and} \quad M = \left(w_i^h, w_j^h\right)$$

are matrices associated with the stiffness and the mass, respectively, and P^h represents the pressure at the computational nodes. Define the computational mesh consisting of nodes and elements containing information on the boundary conditions and medium properties according to the algorithm below:

```
clear;clc;
% Pre-processing: Domain Definition
kappa=1; Xmin=0; Xmax=1;
% Generation of nodes (coordinates)
   nnodes=100; % Assigning nodes numbers
   dx=(Xmax-Xmin)/(nnodes-1);
x=Xmin:dx:Xmax;
x1=Xmin;
x2=Xmax;
% Total number of elements
nele=nnodes-1;
% Conectivity matrix
ien=zeros(nele,2);
for e=1:nele
  ien(e,1)=e;
  ien(e,2)=e+1;
end
% BC, Dirichlet nodes
iccd=[1 nnodes];
uccd=[0   0];
% BC Newmann nodes
iccn=[];
uccn=[];
% Degrees of freedom
nccd=size(iccd,2);
nccn=size(iccn,2);
```

The vector `iccd` contains the nodes indices at the Dirichlet boundary.

Show that transforming the original eigenvalue-eigenfunction problem in a generalized eigenvalue-eigenvector problem with eigenvalues ω^2 and eigenvectors P^h further requires the application of discrete BCs as described below:

Dirichlet boundary conditions are straightly imposed in Finite Elements formulations, being then denoted as *essential*. As a first step, boundary nodes are identified. In one-dimensional problems, boundary nodes are the ones at the domain extremes, $i = 1$ and $i = n$ nodes. For a node i, at the system boundary we have:

(a) Prescribe the condition $P(i) = 0$, by assign zeroes to all elements A and M in line i;
(b) Impose that nodal values also vanish, by doing $A(i, i) = 1$.

Vanishing derivatives boundary conditions are implicitly imposed in Finite Element formulations, being thus denoted as *natural* boundary conditions.

Verify that Dirichlet boundary conditions are properly assigned with the following algorithm, written in *Octave*:

```
for e=1:nele
 lenght=(x(ien(e,2))-x(ien(e,1)));
 k=[1 -1;-1 1]/lenght; % stiffness matrix
 m=[2 1;1 2]*lenght/6; % mass matrix
 for ilocal=1:2
  iglobal=ien(e,ilocal);
  for jlocal=1:2
   jglobal=ien(e,jlocal);
   K(iglobal,jglobal)=K(iglobal,jglobal)+k(ilocal,jlocal);
   M(iglobal,jglobal)=M(iglobal,jglobal)+kappa*m(ilocal,jlocal);
  end
 end
end
```

The indices of the Dirichlet boundary are stored in the vector idbcd.

The following procedure written in *Octave* language illustrates the implementation of BCs to matrices K and M:

```
% setting BCs (zeroing line and setting 1 to main diagonal)
for i=1:nccd
% zeroing line iccd[i]
K(iccd(i),:)=0;
M(iccd(i),:)=0;
% setting 1 to main diagonal
K(iccd(i),iccd(i))=1;
end;

% Solving eigenvalues problem
neigs=5;
[V,d] = eigs(K,M,neigs,1);
sqrt(d/4/pi/pi);
hold off

% Plotting first 5 eigenfunctions
for i=1:neigs
```

```
plot(x,real(V(:,i)),x,imag(V(:,i)));
hold on
end

% Plotting first 4 eigenfunctions
xlabel('domain lenght x','FontSize', 20)
xtickformat('%.1f')
ylabel('amplitude','FontSize', 20)
ytickformat('%.2f')
set(gca,'FontSize',20)
```

Validation Tests: Resonance in a Tube with Open Ends

Validate the above methodology by evaluating the eigenvalues and eigenfunctions associated with a steady wave in a tube with open ends ($p = 0$) to demonstrate the correctness of the method. Show that the eigenfunctions take the form shown in Fig. 1.3.

4. **Formulation of the real eigenvalue-eigenfunction problem in 2D:**
   Consider now a closed two-dimensional cavity and the problem of finding the resonance frequency, near a frequency of interest, $\omega_0 = 2\pi f_0$, and the associated vibration modes, known as standing or stationary waves. Write, in dimensional form, $\kappa^* = 2\pi f_0^*$, where $f_0^* = f/f_0$ is the nondimensional resonant frequency. Show that dropping the asterisks leads to an eigenvalue-eigenfunction problem given by:

$$\nabla^2 p = -\omega^2 p.$$

The eigenvalues $\omega = \kappa$ represent the nondimensional resonant angular frequencies of the cavity, and the associated standing modes are the eigenfunctions of the Laplacian operator. Verify that, at an impermeable boundary, with a local normal unitary vector $\mathbf{n}$, boundary conditions can be written in the form:

$$\mathbf{n} \cdot \nabla p = 0. \tag{1.22}$$

Fig. 1.3 Eigenfunctions associated with a steady wave in a tube with open ends ($p = 0$) to demonstrate the correctness of the method. The result confirms that the eigenfunctions are Fourier modes satisfying the Dirichlet homogeneous boundary conditions

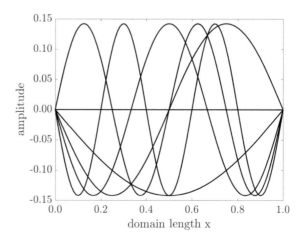

Formulation of the Real Eigenvalue-Eigenfunction Problem in 2D, Through the Finite Elements Method

Discrete formulation of the real eigenvalue-eigenfunction problem in 2D through the Finite Elements Method

The Galerkin formulation of the two-dimensional eigenvalue-eigenfunction problem consists in choosing base and weighing functions in finite dimensional spaces $\mathscr{P}^h$ and $\mathscr{W}^h$. The discrete problem turns out to finding $p^h \in \mathscr{P}^h$ and $\omega \in \mathscr{R}$ is such that, for all $W_h \in \mathscr{W}^h$

$$\left(\nabla p^h, \nabla W^h\right) = -\left(\kappa p^h, \kappa W^h\right).$$

Upon using canonical Finite Element basis in Cartesian coordinates, one obtains the following linear system in matrix form:

$$AP = -\kappa^2 MP,$$

where A and M are the associated rigidity and mass matrices, respectively, whereas P stems for p values at the computational nodes. In order to approximate the original problem to a generalized eigenvalue-eigenvector problem with eigenvalue κ^2 and eigenvector P, complete the problem statement by applying the appropriate boundary conditions. In the case of a closed cavity, BCs consist in prescribing impermeable walls, which are naturally satisfied in the boundaries by the variational formulation.

Computer Implementation of the Finite Elements Scheme

Implementation of the main program and of the auxiliary functions in *Octave* to run the above scheme is suggested below:

Main Program

```
% %% --------------- Main ------------------------------ %%%
% Mesh parameters
Ly=1; %%% vertical size
Lx=2; %%% horizontal size
Nx=21; Ny=11; # number of mesh points
X=zeros(Nx*Ny,1); Y=zeros(Nx*Ny,1);
dx=1/(Nx-1)*Lx; dy=1/(Ny-1)*Ly;
% Mesh generation
[TRI,X,Y] = meshing(Nx,Ny,dx,dy,X,Y);
% Assembling of matrices
[Ak,Mk,Dx,Dy] = assemblingmatrices(TRI,X,Y);
% Solution of the linear system
nmodes=15;
[V,D] = eigs(Ak,Mk,nmodes,'sm');

% Plotting eigenvalue and eigenvector
for imode=2:11 % plotting 10 first modes
 colormap jet
 % eigen plot
 subplot(1,2,1); trisurf(TRI,X,Y,V(:,imode)); shading interp
```

```
title(['Pressure field - f = ' num2str(D(imode,imode))]);
set(get(gca,'title'),'Position',[2.6 1.9])
pbaspect([1 1 1])
xlabel('x','FontSize', 16)
xtickformat('%.1f')
ylabel('y','FontSize', 16)
ytickformat('%.1f')
zlabel('amplitude','FontSize', 16)
ztickformat('%.2f')
set(gca,'FontSize',18)

% quiver plot
u=-Mk\(Dx*V(:,imode));v=-Mk\(Dy*V(:,imode));
subplot(1,2,2);quiver(X,Y,u,v);
title(['Velocity field - f = ' num2str(D(imode,imode))]);
set(get(gca,'title'),'Position',[0.7 1.19])
pbaspect([1 1 1])
xlabel('x','FontSize', 16)
xtickformat('%.1f')
xlim([-0.1 2.1])
ylabel('y','FontSize', 16)
ytickformat('%.1f')
ylim([-0.1 1.1])
set(gca,'FontSize',18)
drawnow
end
% %% --------------- End Main -------------------------- %%%
```

For the Mesh Generation
```
% %% ----------- Creating mesh: X,Y and TRI ------------- %%%
function [TRI,X,Y] = meshing(Nx,Ny,dx,dy,X,Y)
 for i=1:Nx
  for j=1:Ny
   kk=i+(j-1)*Nx;
   X(kk)=(i-1)*dx;
   Y(kk)=(j-1)*dy;
  end
 end
 % setting connectivity matrix TRI
 TRI=zeros((Nx-1)*(Ny-1)*2,3);
 nel=0;
 for i=1:Nx-1;
  for j=1:Ny-1;
   kk=i+(j-1)*Nx;
   nel=nel+1;
   TRI(nel,:)=[kk kk+Nx+1 kk+Nx];
   nel=nel+1;
   TRI(nel,:)=[kk kk+1 kk+Nx+1];
  end
 end
end
% %% --------- End Creating mesh: X,Y and TRI ----------- %%%
```

For Assembling the Matrices

```
% %% ------------ Assembling matrices -------------------- %%%
% builds matrices: stiffness, mass and gradient x and y
function [Ak,Mk,Dx,Dy] = assemblingmatrices(TRI,X,Y)
 numel=size(TRI,1);
 numnode=size(X,1);
 Ak=zeros(numnode,numnode);
 Mk=zeros(numnode,numnode);Dx=Ak;Dy=Ak;
 %%% Mass matrix of element
 Mele = [2 1 1; 1 2 1; 1 1 2];
 %%% Lumped Mass matrix of element
 MeleL = [4 0 0; 0 4 0; 0 0 4];
 for ie=1:numel
  v1=TRI(ie,1); %%% element index 1
  v2=TRI(ie,2); %%% element index 2
  v3=TRI(ie,3); %%% element index 3
  vv=[v1; v2; v3];
  area=0.5*((X(v2)-X(v1))*(Y(v3)-Y(v1))-(X(v3)-X(v1))...
           *(Y(v2)-Y(v1))); %%% element area
  Bt=[Y(v2)-Y(v3)  X(v3)-X(v2); Y(v3)-Y(v1)...
     X(v1)-X(v3); Y(v1)-Y(v2)  X(v2)-X(v1)];
  Aele=Bt*Bt';
  Dxe=1/6*[1; 1; 1]*(Bt(:,1))';
  Dye=1/6*[1; 1; 1]*(Bt(:,2))';
  for i=1:3
   for j=1:3
    Ak(vv(i),vv(j))=Ak(vv(i),vv(j))-Aele(i,j)/area/4;
    Mk(vv(i),vv(j))=Mk(vv(i),vv(j))+Mele(i,j)*area/12;
    Dx(vv(i),vv(j))=Dx(vv(i),vv(j))+Dxe(i,j);
    Dy(vv(i),vv(j))=Dy(vv(i),vv(j))+Dye(i,j);
   end
  end
 end
end
% %% ---------- End Assembling matrices ----------------- %%%
```

Validation Tests: Rectangular Cavity

Apply the above described method to a rectangular cavity with nondimensional $L_x = 2$ and $L_y = 1$, and obtain modes associated with the nondimensional resonant frequency as shown in Fig. 1.4. The figure was obtained in rectangular meshes with 21×11 computational modes showing two one-dimensional and one two-dimensional modes. Left panels show the pressure field and right ones the velocity field. Verify that the results present good agreement with analytical ones.

Pressure field - f = -9.9502   Velocity field - f = -9.9502

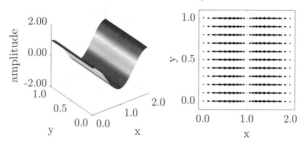

Pressure field - f = -9.9505   Velocity field - f = -9.9505

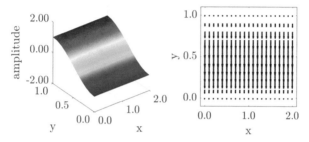

Pressure field - f = -20.2205 Velocity field - f = -20.2205

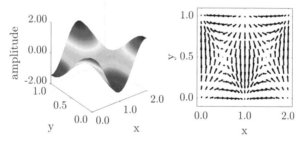

Fig. 1.4 Numerical results in meshes with 21×11 computational nodes, showing two one-dimensional and one two-dimensional modes. Left: pressure, right: velocity field

As a reference, the following table presents the results of simulations conducted in rectangular cavities with dimensions $L_x = 2$ and $L_y = 1$.

n_x	n_y	f_1	f_2	f_3	f_4	f_5
11	6	0.2510	0.5080	0.5081	0.5738	0.7398
21	11	0.2503	0.5020	0.5020	0.5628	0.7157
41	21	0.2501	0.5005	0.5005	0.5600	0.709320
81	41	0.2500	0.5001	0.5001	0.5593	0.7077
161	81	0.2500	0.5000	0.5000	0.5591	0.7072
∞	∞	0.25	0.5	0.5	0.5590	0.7071

5. **Forced solutions of the wave equation** This problem discusses the implementation of the numerical solution of the forced wave equation in one and
in two dimensions, both in the frequency domain and in the time domain,
using the Finite Elements Method. Accordingly, we consider that part of the
domain boundary is excited by a periodic forcing with prescribed amplitude and
frequency:

$$\kappa^2 p + \nabla^2 p = 0,$$

where $\kappa^2 = \omega^2/c_0^2$ stands for the square of the wave number. In this case, the
frequency ω is known and determined by the conditions at forced boundary Γ_f,
which reads, in the time domain:

$$\hat{p}(\Gamma_f, t) = p_0 e^{j\omega t}.$$

The One-Dimensional Forced Problem
Consider a closed cavity except in a region forced with a prescribed frequency
$\omega_0 = 2\pi f_0$. The applicable one-dimensional equation reduces to:

$$\left(\kappa^2 + D^2\right) p = 0,$$

with the forcing condition at the boundary, $p(0) = 1$.

**The Discrete Formulation of the One-Dimensional Problem in the Fre-
quency Domain, with the Finite Elements Method**
Galerkin's formulation of the forced problem consists in choosing base and
weight functions in finite dimensional spaces $\mathscr{E}^h$ and $\mathscr{W}^h$. The discrete problem
reduces to finding $P^h \in \mathscr{P}^h$ and $\omega \in \mathscr{R}$ such that for all $W_h \in \mathscr{W}^h$

$$\left(D P^h, D W^h\right) = -\left(\kappa P^h, \kappa W^h\right).$$

By using canonical basis in Cartesian coordinates, show that we obtain the
following linear system, written in matrix form:

$$A P = -M_\kappa P,$$

where:

$$A = \left(D P^h, D W^h\right); \quad \text{and} \quad M = \left(\kappa P^h, \kappa W^h\right)$$

are the associated rigidity and mass matrices, respectively, and p^h stands for
the pressure in the computational nodes. Note that κ is, in general, a complex
quantity with changing values from an element to other one. Show that the

problem admits a unique solution by prescribing discrete boundary conditions as described below:

Nonreflexive Boundary Conditions
In the case of nonreflexive boundaries we seek solutions of the wave equation propagating in the sense of the external normal direction only, in order to prevent the propagation of waves inside the cavity. Show that this condition is expressed by:

$$\frac{\partial p}{\partial t} + c\mathbf{n} \cdot \nabla p = 0.$$

Show that this condition reads, in the frequency domain:

$$j\omega p + c\mathbf{n} \cdot \nabla p = 0,$$

where c is the wave propagation velocity in the domain. The nonreflexive boundary condition can be generalized for the case of a partially reflexive wall:

$$j\alpha\omega p + c\mathbf{n} \cdot \nabla p = 0.$$

Show that the case $\alpha = 1$ results in the nonreflexive boundary condition, and $\alpha = 0$, in a totally reflexive boundary (rigid wall). Intermediate values of α refer to partially reflexive walls (or partially absorbing ones), also known as mixed or Robin boundary condition. Numerical implementation of this boundary condition can be made in *Octave*, for 2D problems, with one wall perpendicular to the x direction, using the following algorithm:

```
b=zeros(Nx*Ny,1);
b(bn1)=1;
MM=-(Ak+omega2*Mk);
for ii=1:size(bn1)
      i=bn1(ii);
MM(i,:)=0;
MM(i,i)=1;
end
for ii=1:size(bn2,1)
      i=bn2(ii);
MM(i,:)=Dx(i,:)+ Mk(i,:)*alpha*sqrt(-1)*omega;
end
P=MM\b;
end
```

The vector *bn*1 contains the indices of the forced boundary and the vector *bn*2, the indices of the reflexive boundary.

The results of numerical simulations in forced cavities with a partially reflexive boundary in two geometries are presented in Fig. 1.5, as a function of the excitation frequency. The first geometry refers to a rectangular channel with $L_x = 1$ and $L_y = 0.1$. The response is, in this case, essentially the same

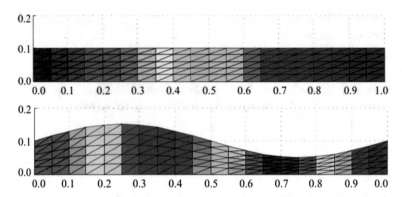

Fig. 1.5 Geometries and numerical meshes used in tests with forcing. The first one refers to a rectangular channel with $Lx = 1$ and $Ly = 0.1$, leading to an essentially one-dimensional response. The second one refers to variable width channel

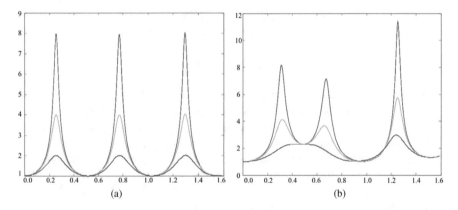

Fig. 1.6 Frequency response of cavities forced at the left boundary, with semi-reflexive boundary at right. Numerical results obtained in meshes with 21×6 computational nodes. (**a**) Result with rectangular geometry. (**b**) Modified geometry. Solid lines, dashed lines, and dot-dashed lines represent the results for a semi-reflexive boundary with $\alpha = 0.125$, 0.250, and 0.5 respectively

obtained in the one-dimensional case. The second geometry refers to a variable width channel produced by a mapping given by $Y = Y.*(1+0.5*\sin(X*2*pi))$.

Figure 1.6 shows that for the rectangular cavity the resonance frequencies are evenly spaced, located at $f = 0.25$, 0.75, and 1.25. The maximum value of the response for all resonance frequencies is $p_{\max} = P_0/\alpha$, where P_0 is the forcing intensity. In the case of the mapped cavity, the resonant frequencies are shifted, and different values of maxima are observed.

6. **Solutions of the wave equation in the time domain**
   Solutions of the forced wave equation in the time domain can be obtained by discretizing the governing equations or by superposition of solutions obtained in the frequency domain.

In order to compare the solutions of the wave with the ones of the Euler equation, consider the following problem in the time domain:

$$\frac{\partial p}{\partial t} + c_0^2 \nabla \cdot (\rho u) = 0 \tag{1.23}$$

$$\frac{\partial \rho u}{\partial t} + \nabla p = 0. \tag{1.24}$$

Show that this system takes the simpler form, for one-dimensional flows:

$$\frac{\partial \mathbf{U}}{\partial t} + \frac{\partial \mathbf{F}}{\partial x} = 0 \qquad \text{in } \Omega \times [0, T_{\max}], \tag{1.25}$$

where $\Omega \subset \mathbf{R}$ and $t \in [0, T_{\max}]$. The vectors of conservative variables $\mathbf{U}$ and of fluxes $\mathbf{F}$ are given by:

$$\mathbf{U} = \begin{Bmatrix} p \\ \rho u \end{Bmatrix}, \qquad \mathbf{F} = \begin{Bmatrix} c_0^2 \rho u \\ p \end{Bmatrix}. \tag{1.26}$$

Show that, as in the one-dimensional case with a closed end at $x = 0$, the boundary conditions reduce to:

$$\frac{\partial p}{\partial x}(x = 0) = 0.$$

If the extreme located at $x = L$ is opened to the atmosphere we can assume $p(L) = 0$. Figure 1.7 shows the solution of the one-dimensional wave equations using finite difference and the MacCormack's method, which will be detailed in the exercise section in Chap. 3 for the solution of one-dimensional hyperbolic conservative equations in the time domain.

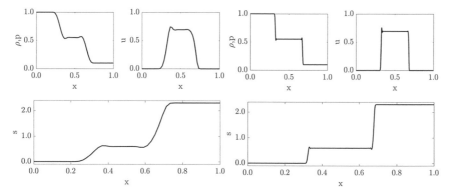

Fig. 1.7 Solution of the pressure and the velocity fields in the wave equation using the MacCormack's method in a uniform grid with 50 nodes (left) and in grid with 1000 points (right). The figure presents a plot of density ρ, velocity u, pressure p, and entropy $s = 1/(\gamma-1)\log(p/\rho^\gamma)$ for each case

7. The electromagnetic field evolves according to the Maxwell equations, given by:

$$\text{div } \mathbf{E} = \frac{\rho}{\varepsilon} \qquad\qquad \text{rot } \mathbf{E} = -\frac{\partial \mathbf{B}}{\partial t}$$

$$\text{div } \mathbf{B} = 0 \qquad\qquad \text{rot } \mathbf{B} = \mu\sigma\mathbf{E} + \mu\varepsilon\frac{\partial \mathbf{E}}{\partial t},$$

where $\mathbf{E}$ and $\mathbf{B}$ represent, respectively, the electric and the magnetic fields, ρ is the local density of electric charges, σ, the electric conductivity, μ and ε, the magnetic permeability and the electric permissively of the medium. Show that the electric and the magnetic fields satisfy the wave equation in the form:

$$\mu\varepsilon\frac{\partial^2 \mathbf{E}}{\partial t^2} + \mu\sigma\frac{\partial \mathbf{E}}{\partial t} = \nabla^2\mathbf{E} - \mathbf{grad}\ (\text{div } \mathbf{E})$$

$$\mu\varepsilon\frac{\partial^2 \mathbf{B}}{\partial t^2} + \mu\sigma\frac{\partial \mathbf{B}}{\partial t} = \nabla^2\mathbf{B},$$

and that the expressions of the electric and the magnetic fields, given by:

$$\mathbf{E}\,(\mathbf{r}, t) = \mathbf{E}_0 \exp\left[i\,(\kappa \cdot \mathbf{r} - \omega t)\right] + cc$$

$$\mathbf{B}\,(\mathbf{r}, t) = \mathbf{B}_0 \exp\left[i\,(\kappa \cdot \mathbf{r} - \omega t)\right] + cc,$$

satisfy the wave equation for mediums without electric charges. In the above equations, $\mathbf{E}_0$ and $\mathbf{B}_0$ are constants, κ and ω stand for the wave vector and the frequency of the propagating electromagnetic perturbation, $\mathbf{r}$ is the position vector, and cc denotes the conjugate of a complex number.

Chapter 2
One-Dimensional Compressible Flows

1 Introduction

This chapter deals with the study of one-dimensional compressible flows [5, 6, 9, 10, 12]. Section 2 addresses isentropic flows in nozzles with variable cross section. We define the Mach number, obtain the area-velocity relation, and equations relating the temperature, pressure, and specific mass at a given point of the nozzle, with the corresponding Mach number at the same point.

Next, we consider the case of strong waves, also including the compression through shock waves, a phenomena comprising entropy production (Sect. 3), and the thickness of shock waves. In sequence, the chapter addresses the effects of friction and heat transfer in variable cross section ducts, where the critical Mach number $M = 1$ is not attained at ducts throat. In sequence, the chapter discusses the flow of gases in isothermal ducts, and the analogies between compressible flows and open channel hydraulics.

2 One-Dimensional Isentropic Gasdynamics

We consider now the one-dimensional steady flow of a perfect gas through convergent-divergent nozzles, as shown in Fig. 2.1. The one-dimensional velocity is defined as the ratio between the volumetric flow rate divided by the local transverse area of the nozzle ($u = Q/A$).

Upon neglecting gravitational and viscous effects, the balance of the convected x-component of the momentum traversing a section with thickness dx perpendicular

Fig. 2.1 The scheme of a
convergent-divergent nozzle

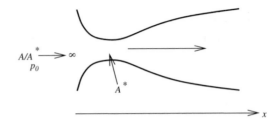

to the flow cross section is obtained from the one-dimensional Euler's equation
(Eq. 5.26), given by:

$$u \, du = -\frac{dp}{\rho}.$$

This equation shows that in the absence of viscous and gravitational effects, the
pressure is always reduced when the velocity increases.

In addition to neglecting viscous and gravitational effects, we also neglect heat
transfer and internal heat generation. In consequence of these assumptions, along
with the hypothesis of inviscid fluid, the flow is isentropic.

$$u \, du = -a^2 \frac{d\rho}{\rho}. \tag{2.1}$$

This relation shows that the specific mass always declines when the velocity
increases. Besides, the larger the local velocity is, the larger will be the specific mass
decrease. Beyond a certain limit velocity, the specific mass decrease is sufficiently
so large that the cross section of the nozzle must increase to cope with the velocity
increases.

Let us express the sound velocity a as a function of other variables of the flow.
For an isentropic flow, we have $p/\rho^\gamma = C^{te}$, where γ is the isentropic coefficient
defined as the ratio between the specific heats at constant pressure, C_p and constant
volume, C_v. We have, thus:

$$p = C^{te} \rho^\gamma$$

$$\left(\frac{\partial p}{\partial \rho}\right)_s = C^{te} \gamma \rho^{\gamma-1} = \frac{p}{\rho^\gamma} \gamma \rho^{\gamma-1} = \gamma \frac{p}{\rho}.$$

Since $p/\rho = RT$, we write:

$$\left(\frac{\partial p}{\partial \rho}\right)_s = a^2 = \gamma \frac{p}{\rho} = \gamma RT. \tag{2.2}$$

The term dp/ρ of Eq. 2.1 can be replaced using the continuity equation, $\rho Au = C^{te}$,
resulting in a relation between the flow velocity and the local area of the nozzle
transverse section. The continuity equation can also be written as:

$$\frac{d\rho}{\rho} + \frac{dA}{A} + \frac{du}{u} = 0,$$

since integration of this last equation results in:

$$\log \rho + \log A + \log u = C^{te}$$

or $\rho A u = C^{te}$. By replacing:

$$\frac{d\rho}{\rho} = -\left(\frac{dA}{A} + \frac{du}{u}\right)$$

in Eq. 2.1 we obtain:

$$u \, du = a^2 \left(\frac{dA}{A} + \frac{du}{u}\right).$$

Rearranging terms, we have:

$$\frac{u}{a^2} du - \frac{du}{u} = \frac{dA}{A}.$$

Mach number is defined as:

$$M = \frac{u}{a}. \tag{2.3}$$

Using this definition and putting the term du/u in evidence we obtain:

$$\left(M^2 - 1\right)\frac{du}{u} = \frac{dA}{A}. \tag{2.4}$$

Several conclusions can be taken from Eq. 2.4: first, the equation shows that for $M < 1$ the coefficient of the term du/u is *negative*. Consequently, the flow velocity increases when the nozzle the cross section decreases. The flow behavior is qualitatively the same that was found in incompressible flows. However, when $M > 1$, the coefficient of the term du/u is positive, showing that the cross section area must increase if so does the velocity. $M = 1$ is the limit value, above which the specific mass decreases faster than the velocity increases, qualitatively changing the flow behavior. Velocities above $M = 1$ can only be obtained with increases in the cross section area. Finally, Eq. 2.4 shows that $M = 1$ is attained in isentropic flows only if the cross section area of the nozzle does not vary, namely if $(dA/A = 0)$, as in nozzle throats. At these points, the flow can either accelerate or decelerate. So, supersonic one-dimensional isentropic flows are attained, starting from subsonic flows, only through convergent-divergent nozzles, with $M = 1$ being attained in the throat. The pressure conditions required for supersonic flow downstream the throat are discussed in the next section.

Nevertheless, it must be pointed that the above discussion refers to conditions to attain reversibly a supersonic one-dimensional flow, starting from subsonic conditions. Supersonic flow may also be attained without convergent-divergent nozzles, for instance, in explosions, where the isentropic condition is not satisfied.

We now relate the temperature, pressure, and specific mass of a one-dimensional isentropic flow with the local Mach number, by making use of the equation of the stagnation enthalpy, h_0, where $h_0 = h + u^2/2$:

$$\frac{Dh_0}{Dt} = \frac{1}{\rho}\frac{\partial p}{\partial t} + \frac{1}{\rho}\frac{\partial}{\partial x_j} v_i \tau_{ij} + v_i g_i + \frac{\kappa}{\rho}\nabla^2 T + \frac{\dot{Q}}{\rho}.$$

Upon assuming steady inviscid flow and neglecting gravitational effects and heat transfer between fluid particles, the equation of the stagnation enthalpy becomes:

$$\frac{Dh_0}{Dt} = 0,$$

namely, $h_0 = C^{te}$. We have, then:

$$h_0 = h + \frac{u^2}{2} = C^{te}.$$

Having in mind that the enthalpy of a perfect gas is given by $h = C_p T$, we write:

$$C_p T_0 = C_p T + \frac{u^2}{2} = C^{te}, \tag{2.5}$$

where T_0 is the *stagnation temperature*, i.e., the flow absolute temperature at rest.

Specific heats are eliminated, reminding that $C_p - C_v = R$:

$$\frac{C_p}{C_v} - 1 = \frac{R}{C_v} \quad \Longrightarrow \quad \frac{R}{C_v} = \gamma - 1.$$

Noting that:

$$a^2 = \gamma RT = \frac{C_p}{C_v} RT \quad \Longrightarrow \quad C_p T = \frac{a^2}{R/C_v} = \frac{a^2}{\gamma - 1}. \tag{2.6}$$

We rewrite Eq. 2.5 as

$$\frac{a_0^2}{\gamma - 1} = \frac{a^2}{\gamma - 1} + \frac{u^2}{2} = C^{te}, \tag{2.7}$$

where a_0 and a are the values of the sound velocity at stagnation and local temperatures, given, respectively, by T_0 and T. Multiplying the above equation by $(\gamma - 1)/a^2$, we have:

$$\frac{a_0^2}{a^2} = 1 + \frac{\gamma - 1}{2}\frac{u^2}{a^2}.$$

Considering that $a_0^2/a^2 = \gamma RT_0/\gamma RT = T_0/T$ and that $u^2/a^2 = M^2$, we obtain an equation relating the stagnation and local temperatures to the associated local Mach number:

$$\frac{T_0}{T} = 1 + \frac{\gamma - 1}{2}M^2. \tag{2.8}$$

Using the relations

$$\frac{\rho_0}{\rho} = \left(\frac{T_0}{T}\right)^{1/(\gamma-1)} \quad \text{and} \quad \frac{p_0}{p} = \left(\frac{T_0}{T}\right)^{\gamma/(\gamma-1)},$$

we have:

$$\frac{\rho_0}{\rho} = \left(1 + \frac{\gamma - 1}{2}M^2\right)^{1/(\gamma-1)} \tag{2.9}$$

and

$$\frac{p_0}{p} = \left(1 + \frac{\gamma - 1}{2}M^2\right)^{\gamma/(\gamma-1)}. \tag{2.10}$$

We denote the flow conditions at the point where the Mach number is equal to one as *critical conditions* and represent variables flow properties at these conditions with an asterisk (ρ^*, p^*, T^*, etc.). Sound velocity at $T = T^*$ is denoted as a^*. The relation between critical and stagnation temperatures, specific masses, and pressures for air ($\gamma = 1, 4$) is obtained with Eqs. 2.8, 2.9, and 2.10.

$$\frac{T^*}{T_0} = \left(1 + \frac{\gamma - 1}{2}\right)^{-1} = 0.833$$

$$\frac{\rho^*}{\rho_0} = \left(1 + \frac{\gamma - 1}{2}\right)^{-1/(\gamma-1)} = 0.634$$

$$\frac{p^*}{p_0} = \left(1 + \frac{\gamma - 1}{2}\right)^{-\gamma/(\gamma-1)} = 0.528.$$

We derive now the relation between the transversal area of the nozzle at a point where the Mach number is M, and the critical area where $M = 1$. Now, we consider the relationship between the area of some section where the Mach number is M and the critical area of the nozzle as function of M, i.e., we derive a relation in the form $A/A^* = f(M)$. Accordingly, we use the continuity equation in steady one-dimensional, $\rho A u = C^{te}$. In particular:

$$\rho A u = \rho^* A^* u^*.$$

This equation can be written as:

$$\frac{A}{A^*} = \frac{\rho^*}{\rho} \frac{a^*}{u} = \frac{\rho^*}{\rho_0} \frac{\rho_0}{\rho} \frac{1}{M^*},$$

where M^* is the Mach number defined as the ratio between the local velocity and the sound velocity where $u = a$.

The relations ρ^*/ρ_0 and ρ_0/ρ are obtained from Eq. 2.9:

$$\frac{\rho^*}{\rho_0} = \left(1 + \frac{\gamma - 1}{2}\right)^{-1/(\gamma-1)} = \left(\frac{\gamma + 1}{2}\right)^{-1/(\gamma-1)} = \left(\frac{2}{\gamma + 1}\right)^{1/(\gamma-1)}$$

$$\frac{\rho_0}{\rho} = \left(1 + \frac{\gamma - 1}{2} M^2\right)^{1/(\gamma-1)} .$$

So:

$$\frac{\rho^*}{\rho_0} \frac{\rho_0}{\rho} = \left[\frac{2}{\gamma + 1}\left(1 + \frac{\gamma - 1}{2}\right) M^2\right]^{1/(\gamma-1)}$$

and

$$\left(\frac{A}{A^*}\right) = \left[\frac{2}{\gamma + 1}\left(1 + \frac{\gamma - 1}{2} M^2\right)\right]^{1/(\gamma-1)} \frac{1}{M^*}. \qquad (2.11)$$

We express now $1/M^{*2} = f(M)$ using Eq. 2.7 to write:

$$\frac{a_0^2}{\gamma - 1} = \frac{a^2}{\gamma - 1} + \frac{u^2}{2} = \frac{a^{*2}}{\gamma - 1} + \frac{a^{*2}}{2},$$

or

$$\frac{a^2}{\gamma - 1} + \frac{u^2}{2} = \frac{\gamma + 1}{2(\gamma - 1)} a^{*2}.$$

Upon dividing this last equation by u^2, we have:

$$\frac{1}{M^2}\frac{1}{\gamma-1}+\frac{1}{2}=\frac{\gamma+1}{2(\gamma-1)}\frac{1}{M^{*2}},$$

what leads to:

$$\frac{1}{M^{*2}}=\left(\frac{1}{M^2}\frac{1}{\gamma-1}+\frac{1}{2}\right)\frac{2(\gamma-1)}{\gamma+1}=\frac{2}{M^2(\gamma+1)}+\frac{\gamma-1}{\gamma+1}$$

$$=\frac{2+(\gamma-1)M^2}{(\gamma+1)M^2}=\frac{1+\dfrac{\gamma-1}{2}M^2}{\dfrac{\gamma+1}{2}M^2}.$$

From the above equation we conclude that:

$$\frac{1}{M^{*2}}=\frac{2}{\gamma+1}\left(1+\frac{\gamma-1}{2}M^2\right)\frac{1}{M^2}. \tag{2.12}$$

Introducing the above result in Eq. 2.11, we find:

$$\left(\frac{A}{A^*}\right)^2=\left[\frac{2}{\gamma+1}\left(1+\frac{\gamma-1}{2}M^2\right)\right]^{2/(\gamma-1)}\frac{2}{\gamma+1}\left(1+\frac{\gamma-1}{2}M^2\right)\frac{1}{M^2}.$$

So:

$$\left(\frac{A}{A^*}\right)^2=\frac{1}{M^2}\left[\frac{2}{\gamma+1}\left(1+\frac{\gamma-1}{2}M^2\right)\right]^{(\gamma+1)/(\gamma-1)}.$$

This equation presents the sought relation between the area of the transverse section where the Mach number is M, with the throat area of the nozzle, where $M=1$. Figure 2.2 schematically shows the profiles p/p_0 and T/T_0 along a nozzle where flow evolves from subsonic to supersonic conditions.

It should be noted that a very well-defined value of the pressure at the nozzle outlet must be settled to assure isentropic flow along the entire nozzle length. This distinguished pressure value is significantly smaller than the outflow pressure required for subsonic flow along the entire nozzle, but with critical conditions at the throat.

Should an intermediate value between the above ones be prescribed for the outflow pressure, no isentropic solution for the flow exists. In fact a steady shock wave settles in the divergent part of the nozzle, as shown schematically in Fig. 2.2c. The flow returns to the subsonic regime downstream the shock wave; the velocity diminishes and pressure rises to attain the prescribed intermediate value at the nozzle exit. The shock is an irreversible process, with entropy production (see Sect. 3.3).

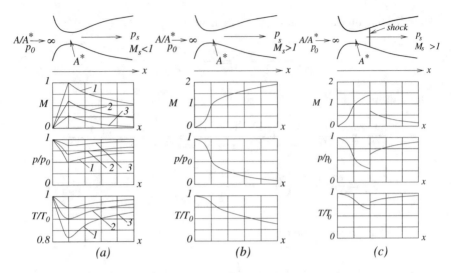

Fig. 2.2 Schematic representation of the Mach number, pressure, and temperatures along a convergent-divergent nozzle operating in the subsonic (**a**) and supersonic regimes (**b**), with isentropic flow throughout in both cases. Mach number, pressure, and temperatures' distribution (**c**), with a localized shock wave in the divergent part of the nozzle (diagrams not in scale)

2.1 Effect of the Mach Number on Flow Parameters

This section discusses the effect of the Mach number on variations of the gas thermodynamic properties, namely variations of the pressure, temperature, and specific mass, in response to variations of the flow velocity. Isentropic, inviscid flow is assumed throughout.

From Euler's equation (Eq. 5.26), we can write:

$$\frac{dp}{p} = -\frac{\rho u^2}{p}\frac{du}{u}.$$

Introducing the expression of the sound velocity in the above equation, we find:

$$\frac{dp}{p} = -\gamma\frac{\rho u^2}{a^2}\frac{du}{u},$$

and:

$$\frac{dp}{p} = -\gamma M^2\frac{du}{u}. \tag{2.13}$$

The effect of velocity variations on the gas temperature is evaluated using the energy equation (Eq. 5.29):

$$\frac{dT}{T} = -\frac{u^2}{C_p T}\frac{du}{u},$$

which can be rewritten, using the expression of the sound velocity, as:

$$\frac{dT}{T} = -\frac{\gamma R}{C_p}M^2\frac{du}{u}.$$

Introducing the relation $R/C_p = \gamma - 1$, in the above equation, we find:

$$\frac{dT}{T} = -(\gamma - 1)\, M^2\frac{du}{u}. \tag{2.14}$$

To find the effect of velocity variations on the gas specific mass, we use the perfect gas equation, written in the form:

$$\frac{dp}{p} = \frac{d\rho}{\rho} + \frac{dT}{T}.$$

Introducing the expressions of pressure and temperature variations, given by Eqs. 2.13 and 2.14, we have:

$$\frac{d\rho}{\rho} = -\gamma M^2\frac{du}{u} + (\gamma - 1)\, M^2\frac{du}{u},$$

what leads to, after rearranging terms:

$$\frac{d\rho}{\rho} = -M^2\frac{du}{u}. \tag{2.15}$$

Equations 2.13, 2.14, and 2.15 show that the flow experiences, respectively, larger pressure, temperature, and specific mass variations for a given velocity variation, at higher Mach numbers, than at lower ones.

2.2   The Pressure Coefficient

In this section we discuss some properties of the dynamic pressure in compressible flows, and define the Pressure Coefficient. By combining the energy equation for adiabatic flow, expressed by Eq. 2.5 with Eq. 2.6, and using the law of perfect gases $p = \rho RT$ to eliminate the temperature T, we write:

$$\frac{1}{2}u^2 + \frac{\gamma}{\gamma - 1}\frac{p}{\rho} = \frac{\gamma}{\gamma - 1}\frac{p_0}{\rho_0}.$$

Using the condition of isentropic process, expressed by $p/\rho^\gamma = p_0/\rho_0^\gamma$, we eliminate the specific mass ρ, to obtain:

$$\frac{1}{2}u^2 + \frac{\gamma}{\gamma - 1}\frac{p_0}{\rho_0}\left(\frac{p}{p_0}\right)^{\gamma - 1/\gamma} = \frac{\gamma}{\gamma - 1}\frac{p_0}{\rho_0}.$$

In compressible flows the dynamic pressure, $\rho U^2/2$, no longer represents the difference between stagnation and static pressures, as in the case of incompressible flows. In addition to the static pressure, it also depends on the flow Mach number. For a perfect gas, this dependence appears as shown below:

$$\frac{1}{2}\rho U^2 = \frac{1}{2}a^2 M^2 = \frac{1}{2}\rho\left(\frac{\gamma p}{\rho}\right)M^2.$$

The dynamic pressure is used as the normalization factor used in the definition of the Pressure Coefficient, C_p:

$$C_p = \frac{p - p_\infty}{\rho U^2/2} = \frac{2}{\gamma M_\infty^2}\frac{p - p_\infty}{p_\infty}. \tag{2.16}$$

For perfect gases, we have:

$$C_p = \frac{p - p_\infty}{\rho U^2/2} = \frac{p - p_\infty}{\gamma p_\infty M_\infty^2} = \frac{2}{\gamma M_\infty^2}\left(\frac{p}{p_\infty} - 1\right),$$

where U, p_∞ and M_∞ are reference values of the velocity, pressure and Mach number, respectively. Using Eq. 2.10 we rewrite the above equation in terms of the reference and local Mach numbers:

$$C_p = \frac{2}{\gamma M_\infty^2}\left\{\left[\frac{2 + (\gamma - 1) M_\infty^2}{2 + (\gamma - 1) M^2}\right]^{\gamma/(\gamma - 1)} - 1\right\},$$

where M is the local Mach number. From the definition of Mach number, $M_\infty = U/a_\infty$, $M = u/a$, and the energy equation in the form:

$$\frac{u^2}{2} + \frac{a^2}{\gamma - 1} = \frac{U^2}{2} + \frac{a_\infty^2}{\gamma - 1},$$

we eliminate the local sound velocity a to obtain:

$$C_p = \frac{2}{\gamma M_\infty^2}\left\{\left[1 + \frac{\gamma - 1}{2}M_\infty^2\left(1 - \frac{u^2}{2}\right)\right]^{\gamma/(\gamma - 1)} - 1\right\}. \tag{2.17}$$

3 Strong Waves: Shock Compression

3.1 Basic Equations

This section deals with the compression of a gas flow crossing a steady shock wave. We prove that the process is irreversible, with entropy production as the gas crosses the shock wave, what defines the direction of the shock process.

   We consider a one-dimensional shock wave traversed by a perfect gas with gas constant R (see Fig. 2.3). By crossing the wave, the gas undergoes a velocity jump from u_1 to u_2, whereas the specific mass and the pressure undergo ρ_1 to ρ_2 and p_1 to p_2 jumps, respectively. Since the shock thickness is small when compared to the duct or body characteristic dimension where the shock occurs, we can assume the area of the transverse section as being constant. In these conditions, the steady one-dimensional continuity, momentum, and energy equations in integral form, neglecting gravitational and viscous effects (Eqs. 5.16, 5.27, and 5.29), take the following form, when applied to a control volume with constant transversal section:

$$\rho_1 u_1 = \rho_2 u_2 \tag{2.18}$$

$$p_1 + \rho_1 u_1^2 = p_2 + \rho_2 u_2^2 \tag{2.19}$$

$$h_1 + \frac{u_1^2}{2} = h_2 + \frac{u_2^2}{2}. \tag{2.20}$$

3.2 The Rankine-Hugoniot Relation

The enthalpy of a perfect gas can be written in the form:

$$h = e + \frac{p}{\rho} = C_v T + \frac{p}{\rho} = C_v \frac{p}{\rho R} + \frac{p}{\rho} = \frac{p}{\rho}\left(\frac{C_v}{R} + 1\right)$$

$$= \frac{p}{\rho}\frac{C_p}{C_p - C_v} = \frac{p}{\rho}\frac{\gamma}{\gamma - 1}.$$

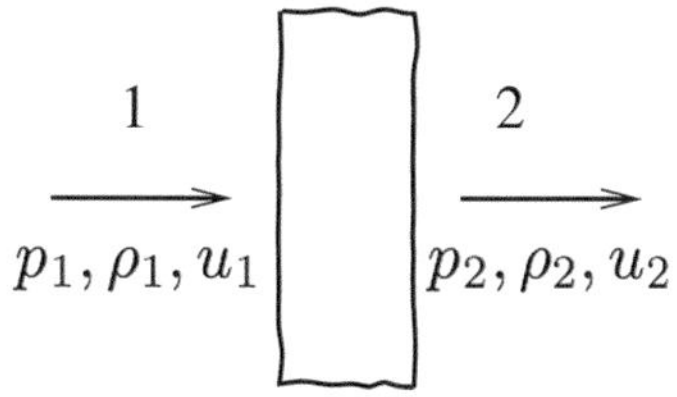

Fig. 2.3 The one-dimensional flow of a perfect gas through a shock wave with constant transversal section

Introducing this result in Eq. 2.20 we obtain:

$$\frac{\gamma}{\gamma - 1} \frac{p_1}{\rho_1} + \frac{u_1^2}{2} = \frac{\gamma}{\gamma - 1} \frac{p_2}{\rho_2} + \frac{u_2^2}{2}, \tag{2.21}$$

which can be rewritten as:

$$\frac{2\gamma}{\gamma - 1} \frac{p_1}{\rho_1} + \frac{\rho_1 u_1^2}{\rho_1} = \frac{2\gamma}{\gamma - 1} \frac{p_2}{\rho_2} + \frac{\rho_2 u_2^2}{\rho_2}.$$

Rearranging terms:

$$\frac{\rho_2}{\rho_1} \left[\frac{2\gamma}{\gamma - 1} p_1 + \rho_1 u_1^2 \right] = \frac{2\gamma}{\gamma - 1} p_2 + \rho_2 u_2^2.$$

Dividing both members of the above equation by p_1, we have:

$$\frac{\rho_2}{\rho_1} \left[\frac{2\gamma}{\gamma - 1} + \frac{\rho_1 u_1^2}{p_1} \right] = \frac{2\gamma}{\gamma - 1} \frac{p_2}{p_1} + \frac{\rho_2 u_2^2}{p_1}. \tag{2.22}$$

From Eq. 2.19:

$$\frac{\rho_1 u_1^2}{p_1} = \frac{p_2}{p_1} + \frac{\rho_2 u_2^2}{p_1} - 1 \qquad\qquad \frac{\rho_2 u_2^2}{p_1} = 1 - \frac{p_2}{p_1} + \frac{\rho_1 u_1^2}{p_1}.$$

Replacing the above results in Eq. 2.22, we find:

$$\frac{\rho_2}{\rho_1} \left[\frac{2\gamma}{\gamma - 1} - 1 + \frac{p_2}{p_1} + \frac{\rho_2 u_2^2}{p_1} \right] = \frac{2\gamma}{\gamma - 1} \frac{p_2}{p_1} - \frac{p_2}{p_1} + 1 + \frac{\rho_1 u_1^2}{p_1}.$$

Rearranging terms once more:

$$\frac{\rho_2}{\rho_1} \left[\frac{\gamma + 1}{\gamma - 1} + \frac{p_2}{p_1} \right] = \frac{\gamma + 1}{\gamma - 1} \frac{p_2}{p_1} + 1 - \frac{\rho_2^2 u_2^2}{\rho_1 p_1} + \frac{\rho_1 u_1^2}{p_1}.$$

Upon noting that Eq. 2.18 shows that $\rho_2^2 u_2^2 = \rho_1^2 u_1^2$, the two last terms of the above equation are equal to zero. We obtain the Rankine-Hugoniot relation, describing the compression process undergone by a perfect gas traversing a shock wave:

$$\frac{\rho_2}{\rho_1} = \frac{u_1}{u_2} = \frac{\dfrac{\gamma + 1}{\gamma - 1} \dfrac{p_1}{p_2} + 1}{\dfrac{\gamma + 1}{\gamma - 1} + \dfrac{p_1}{p_2}}. \tag{2.23}$$

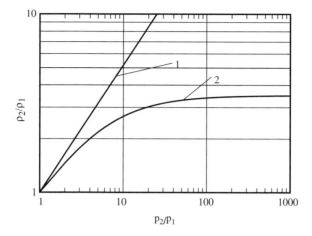

Fig. 2.4 Comparison between the increase of the specific mass of the fluid in isentropic compression (curve N°. 1) and across a shock wave (curve N°. 2), as given by Eq. 2.23

This equation can be written as a relation between the gas temperatures immediately up and downstream the shock wave, assuming the perfect gas law. We obtain:

$$\frac{T_2}{T_1} = \frac{p_2}{p_1} \frac{\dfrac{\gamma + 1}{\gamma - 1} + \dfrac{p_2}{p_1}}{1 + \dfrac{\gamma + 1}{\gamma - 1} \dfrac{p_1}{p_2}}. \tag{2.24}$$

The qualitative difference between an irreversible compression through a shock wave and an isentropic compression between the same initial and final pressures is shown in Fig. 2.4. For pressure ratios smaller than two, variations of the specific mass are approximately the same in both cases. For higher values of the pressure ratio, as found in explosions, for instance, the behavior of the specific mass and, consequently, of the entropy diverges from the one found in isentropic compression. The point is further discussed in Sect. 3.3.

3.3 Entropy Production and Irreversibility of Shock Waves I

From Eq. 2.19, we obtain:

$$\rho_1 u_1^2 - \rho_2 u_2^2 = p_2 - p_1.$$

Taking into account that $\rho_1 u_1 = \rho_2 u_2$, we have from the above equation:

$$u_2 - u_1 = \frac{p_2}{\rho_2 u_2} - \frac{p_1}{\rho_1 u_1}.$$

Since $p/\rho = a^2/\gamma$, according to Eq. 2.2, we can write:

$$u_2 - u_1 = \frac{a_2^2}{\gamma u_2} - \frac{a_1^2}{\gamma u_1}. \tag{2.25}$$

Equation 2.21 can be rewritten, taking into account Eq. 2.2 and the definition of a^*, in the form:

$$\frac{u_1^2}{2} + \frac{a_1^2}{\gamma - 1} = \frac{u_2^2}{2} + \frac{a_2^2}{\gamma - 1} = \frac{1}{2}\frac{\gamma + 1}{\gamma - 1}a^*, \tag{2.26}$$

from where we obtain:

$$u_1 = \frac{\gamma + 1}{\gamma - 1}\frac{a^{*2}}{u_1} - \frac{a_1^2}{u_1}\frac{2}{\gamma - 1} \qquad \text{and} \qquad u_2 = \frac{\gamma + 1}{\gamma - 1}\frac{a^{*2}}{u_2} - \frac{a_2^2}{u_2}\frac{2}{\gamma - 1}.$$

Subtracting the second equation from the above first one, and multiplying the result by γ, we find:

$$\frac{u_2 - u_1}{\gamma} = \frac{\gamma + 1}{\gamma(\gamma - 1)}a^{*2}\frac{u_2 - u_1}{u_1 u_2} - \frac{2}{\gamma - 1}\left(\frac{a_1^2}{\gamma u_1} - \frac{a_2^2}{\gamma u_2}\right).$$

Considering Eq. 2.25, we rewrite:

$$\frac{\gamma + 1}{\gamma(\gamma - 1)}a^{*2}\frac{u_2 - u_1}{u_1 u_2} = -\frac{u_2 - u_1}{\gamma} - \frac{2}{\gamma - 1}(u_1 - u_2)$$

$$= \frac{\gamma + 1}{\gamma(\gamma - 1)}(u_2 - u_1),$$

and conclude that:

$$u_1 u_2 = a^{*2} \qquad\qquad \text{(Prandtl-Meyer relation)}. \tag{2.27}$$

From Eq. 2.12, we have:

$$M^{*2} = \frac{(\gamma + 1)M^2}{2 + (\gamma - 1)M^2}. \tag{2.28}$$

From this equation we find that, in particular:

$$\lim_{M \to \infty} M^* = \left(\frac{\gamma + 1}{\gamma - 1}\right)^{1/2}. \tag{2.29}$$

If we consider flow of air with $\gamma = 1.4$, $M^* = 2.45$ when $M \rightarrow \infty$. The ratio between velocities up and downstream the shock wave can be written, using the Prandtl-Meyer relation:

$$\frac{u_1}{u_2} = \frac{u_1^2}{u_1 u_2} = \frac{u_1^2}{a^{*2}} = M_1^{*2}.$$

From the Prandtl-Meyer, we obtain:

$$\frac{u_1}{a^*}\frac{u_2}{a^*} = M_1^* M_2^* = 1, \tag{2.30}$$

showing that if $M_1^* > 1$, we necessarily have $M_2^* < 1$ and vice versa. We stress that $M_1^* > 1$ implies in $M_1 > 1$ and that $M_2^* < 1$ implies in $M_2 < 1$. So, if the flow is subsonic in one of the sides of a shock wave, it will be supersonic in the other.

We obtain the ratio between the Mach number at the two sides of a shock wave introducing the expression of M_1^{*2} and of M_1^{*2}, obtained from Eq. 2.12, in Eq. 2.30:

$$\frac{(\gamma + 1) M_1^2}{2 + (\gamma - 1) M_1^2} = \frac{(\gamma + 1) M_2^2}{2 + (\gamma - 1) M_2^2}.$$

From this equation, we have:

$$(\gamma + 1)^2 M_1^2 M_2^2 = \left[2 + (\gamma - 1) M_1^2\right]\left[2 + (\gamma - 1) M_2^2\right]$$

and the expression of the Mach number downstream the shock:

$$M_2^2 = \frac{2 + (\gamma - 1) M_1^2}{2\gamma M_1^2 - \gamma + 1} \tag{2.31}$$

the ratio between the specific masses up and downstream the shock wave is obtained from the ratio between the velocities, also considering the continuity equation:

$$\frac{u_1}{u_2} = \frac{\rho_2}{\rho_1} = \frac{(\gamma + 1)M_1^2}{2 + (\gamma - 1)M_1^2}. \tag{2.32}$$

The pressure ratio is obtained from the continuity and momentum equations (Eqs. 2.18 and 2.19):

$$p_2 - p_1 = \rho_1 u_1^2 - \rho_2 u_2^2 = \rho_1 u_1 (u_1 - u_2).$$

We obtain an expression for the shock intensity rewriting the above equation in nondimensional form:

$$\frac{p_2 - p_1}{p_1} = \frac{\Delta p}{p_1} = \frac{\rho_1 u_1^2}{p_1}\left(1 - \frac{u_2}{u_1}\right).$$

Reminding that $a_1^2 = \gamma p_1/\rho_1$ and using Eq. 2.32 we have:

$$\frac{\Delta p}{p_1} = \gamma \frac{u_1^2}{a_1^2}\left[1 - \frac{(\gamma-1)M_1^2 + 2}{(\gamma+1)M_1^2}\right] = \frac{2\gamma}{\gamma+1}\left(M_1^2 - 1\right), \tag{2.33}$$

which leads to:

$$\frac{p_2}{p_1} = 1 + \frac{2\gamma}{\gamma+1}\left(M_1^2 - 1\right). \tag{2.34}$$

We derive an expression for the temperature ratio at the two sides of the shock wave noting that $T_2/T_1 = (p_2/p_1)(\rho_1/\rho_2)$. We have, then:

$$\frac{T_2}{T_1} = 1 + \frac{2(\gamma-1)}{(\gamma+1)^2}\frac{\gamma M_1^2 + 1}{M_1^2}\left(M_1^2 - 1\right). \tag{2.35}$$

The entropy jump across the shock wave can be expressed as a function of the pressure and specific mass jumps. The entropy variation of perfect gas is given by:

$$\frac{s_2 - s_1}{R} = \ln\left[\left(\frac{p_2}{p_1}\right)^{1/(\gamma-1)}\left(\frac{\rho_2}{\rho_1}\right)^{-\gamma/(\gamma-1)}\right].$$

Introducing Eqs. 2.34 and 2.32 in the above equation, we find:

$$\frac{s_2 - s_1}{R} = \ln\left\{\left[1 + \frac{2\gamma}{\gamma+1}\left(M_1^2 - 1\right)\right]^{1/(\gamma-1)}\left[\frac{(\gamma+1)M_1^2}{2 + (\gamma-1)M_1^2}\right]^{-\gamma/(\gamma-1)}\right\} \tag{2.36}$$

Upon defining $M_1^2 - 1 = m$, we rewrite the above equation as:

$$\frac{s_2 - s_1}{R} = \ln\left[\left(1 + \frac{2\gamma}{\gamma+1}m\right)^{1/(\gamma-1)}(1+m)^{-\gamma/(\gamma-1)}\left(\frac{\gamma-1}{\gamma+1}m + 1\right)^{\gamma/(\gamma-1)}\right].$$

We can simplify this equation for the case where $M_1 \approx 1$, noting that terms between brackets are written in the form $1 + b\varepsilon$, with $\varepsilon \ll 1$. Remembering that $\ln(1 + \varepsilon) = \varepsilon - \varepsilon^2/2 + \varepsilon^3/3 + \ldots$, we identify the terms in ε, ε^2, ε^3, etc. The coefficient multiplying terms in ε and in ε^2 vanish, which leads to:

$$\frac{s_2 - s_1}{R} = \frac{2\gamma}{(\gamma + 1)^2} \frac{m^3}{3} + \text{higher order terms.}$$

We can rewrite this equation in the form:

$$\frac{s_2 - s_1}{R} \approx \frac{2\gamma}{(\gamma + 1)^2} \frac{\left(M_1^2 - 1\right)^3}{3}.$$

Since the entropy cannot decrease in the adiabatic process across a shock wave, it is necessary that $M_1 > 1$, namely, the flow through a shock wave occurs from the supersonic regime upstream the shock, to the subsonic one. We also note that the entropy production across the shock is of third order with respect to the square of the upstream Mach number, being thus small if M_1 is not much bigger than one.

We can also express the entropy jump in terms of the shock pressure jump, using Eq. 2.34. We obtain:

$$\frac{s_2 - s_1}{R} \approx \frac{\gamma + 1}{12\gamma^2} \left(\frac{\Delta p}{p_1}\right)^3. \tag{2.37}$$

So, small pressure jumps across the shock result in velocity and specific mass jumps of same order of magnitude, and in entropy jumps of third order with respect to the former variables. Weak shocks are thus, almost isentropic.

3.4 Entropy Production and Irreversibility of Shock Waves II

This section presents an alternative approach for the evaluation of entropy production across a weak shock wave [5]. The entropy variation across the shock being of third order with respect to the pressure rise suggests that the former could not rise monotonically but instead, present a maximum inside the shock wave, what effectively occurs, as shown in Sect. 3.5.

We define $j = \rho u$, and the specific volume $V = 1/\rho$. Then, we have:

$$u_1 = jV_1 \qquad\qquad u_2 = jV_2,$$

where the subscripts 1 and 2 refer to conditions up and downstream the shock. The continuity, momentum, and energy balances across a weak shock wave are:

$$u_1 V_1 = u_2 V_2$$
$$p_1 + j^2 V_1 = p_2 + j^2 V_2$$
$$h_1 + \frac{u_1^2}{2} = h_2 + \frac{u_2^2}{2}.$$

Noting that:

$$u_2 - u_1 = j\,(V_1 - V_2),\qquad\qquad(2.38)$$

and that, from the momentum equation:

$$p_2 - p_1 = j^2\,(V_1 - V_2)\qquad\text{and}\qquad j^2 = \frac{p_2 - p_1}{V_1 - V_2},$$

we replace the expression for j^2 in Eq. 2.38 to obtain:

$$u_1 - u_2 = \sqrt{(V_1 - V_2)\,(p_2 - p_1)}.\qquad\qquad(2.39)$$

The energy equation is rewritten successively as:

$$h_1 + \frac{1}{2}j^2 V_1^2 = h_2 + \frac{1}{2}j^2 V_2,$$

and:

$$h_1 - h2 + j^2\frac{1}{2}\left(V_1^2 - V_2^2\right).$$

Replacing the expression for j^2 in the above equation, we find:

$$h_1 - h_2 + \frac{1}{2}\,(V_1 + V_2)\,(p_2 - p_1) = 0.\qquad\qquad(2.40)$$

We develop now the term $(h_2 - h_1) = -(h_1 - h_2)$ in powers of $(p_2 - p_1)$ and $(s_2 - s_1)$. Since it will be shown that the entropy production is of order of $\mathcal{O}(\Delta p)^3$, we drop terms with the square and the cube of $(s_2 - s_1)$.

$$h_2 - h_1 = \left(\frac{\partial h}{\partial s_1}\right)_p (s_2 - s_1) + \left(\frac{\partial h}{\partial p_1}\right)_s (p_2 - p_1)$$

$$+ \frac{1}{2}\left(\frac{\partial^2 h}{\partial p_1^2}\right)_s (p_2 - p_1)^2 + \frac{1}{6}\left(\frac{\partial^3 h}{\partial p_1^3}\right)_s (p_2 - p_1)^3.$$

Using the thermodynamic relations:

$$\left(\frac{\partial h}{\partial s}\right)_p = T\qquad\text{and}\qquad \left(\frac{\partial h}{\partial p}\right)_s = V,$$

we obtain:

$$h_2 - h_1 = T_1 (s_2 - s_1) + V_1 (p_2 - p_1)$$

$$+ \frac{1}{2} \left(\frac{\partial V}{\partial p_1} \right)_s (p_2 - p_1)^2 + \frac{1}{6} \left(\frac{\partial^2 V}{\partial p_1^2} \right)_s (p_2 - p_1)^3 . \qquad (2.41)$$

An expression for $(V_1 + V_2)$ is obtained from the development below:

$$V_1 + V_2 = 2V_1 + \left(\frac{\partial V}{\partial p_1} \right)_s (p_2 - p_1) + \frac{1}{2} \left(\frac{\partial^2 V}{\partial p_1^2} \right)_s (p_2 - p_1)^2$$

$$+ \left(\frac{\partial V}{\partial s_1} \right)_p (s_2 - s_1) + \frac{1}{2} \left(\frac{\partial^2 V}{\partial s_1^2} \right)_p (s_2 - s_1)^2 . \qquad (2.42)$$

We replace now the expressions of $h_1 - h_2$, given by Eq. 2.41, and of $(V_1 + V_2)$, given by Eq. 2.42 into Eq. 2.40, to obtain:

$$- \left[T_1 (s_2 - s_1) + V_1 (p_2 - p_1) + \frac{1}{2} \left(\frac{\partial V}{\partial p_1} \right)_s (p_2 - p_1)^2 \right.$$

$$\left. + \frac{1}{6} \left(\frac{\partial^2 V}{\partial p_1^2} \right)_s (p_2 - p_1)^3 \right]$$

$$+ \frac{1}{2} \left[2V_1 + \left(\frac{\partial V}{\partial p_1} \right)_s (p_2 - p_1) + \frac{1}{2} \left(\frac{\partial^2 V}{\partial p_1^2} \right)_s (p_2 - p_1)^2 + \left(\frac{\partial V}{\partial s_1} \right)_p (s_2 - s_1) \right.$$

$$\left. + \frac{1}{2} \left(\frac{\partial^2 V}{\partial s_1^2} \right)_p (s_2 - s_1)^2 \right] (p_2 - p_1) = 0.$$

Simplifying the above equation we obtain another one not containing terms in V_1, V_2, nor in $(\partial V/\partial p)_s (p_2 - p_1)$:

$$\left[T_1 - \frac{1}{2} \left(\frac{\partial V}{\partial s_1} \right)_p (p_2 - p_1) - \frac{1}{4} \left(\frac{\partial^2 V}{\partial s_1^2} \right)_p (s_2 - s_1) (p_2 - p_1) \right] (s_2 - s_1)$$

$$= \frac{1}{12} \left(\frac{\partial^2 V}{\partial p_1^2} \right)_s (p_2 - p_1)^3 .$$

The second and the third terms between braces in the last equation can be dropped since we assume weak shocks, where the two terms are smaller than T_1. In consequence:

$$s_2 - s_1 = \frac{1}{12\,T_1} \left(\frac{\partial^2 V}{\partial p_1^2} \right)_s (p_2 - p_1)^3 , \qquad (2.43)$$

confirming that entropy production across weak shocks is of order of $\mathcal{O}(\Delta p)^3$.

The above result refers to the entropy rise between points far up and downstream the shock. The cubic dependence suggests that the entropy inside the shock wave can have a maximum, and that differences between values of the entropy far inside of the shock can be of order of $(\Delta p)^2$, instead of $(\Delta p)^3$, what actually occurs (see Sect. 3.5 where we address the behavior of variables inside the shock wave).

3.5 The Thickness of Shock Waves

In the preceding sections we have addressed the study of shock waves, assumed as discontinuity surfaces of zero thickness. In this section we consider weak shocks, and take viscous effects into account to estimate the thickness of shock. Viscosity and thermal conductivity are assumed as constant, and the thickness of the shock is assumed to be of order $\mathcal{O}(\delta)$, where δ is a number sufficiently large; so numbers between zero and $-\delta/2$ and $\delta/2$ are suitable for use as coordinates of flow properties along the shock. By considering weak shocks, we assume that $p - p_1$ is a small number of order of $\mathcal{O}(1/\delta)$ Here, p is the pressure in a point inside the shock and p_1, the upstream pressure, far before the shock.

Assuming that the shock cross section is constant (see Fig. 2.3), the continuity equation reads:

$$\rho u = ju = \text{Constant.}$$

For the momentum equation we have to consider viscous effects. Balance of momentum transfer and forces actuating on the flow element leads to:

$$p + \rho u^2 - (2\mu + 3\lambda)\frac{du}{dx} = \text{Constant.}$$

At this point we introduce the specific volume of the gas, $V = 1/\rho$. In consequence, $u = jV$, and, using the continuity equation, $du/dx = dV/dx$. The momentum equation becomes:

$$p + j^2 V - (2\mu + 3\lambda)\frac{dV}{dx} = \text{Constant.} \qquad (2.44)$$

Far from the shock waves the variables are independent of x, and, in particular, $du/dx = dV/x = 0$. The value of the constant in Eq. 2.44 is thus $p_1 + j^2 V$. The momentum equation applied between points far upstream and inside the shock wave becomes:

$$p - p_1 + j^2 (V - V_1) - j (2\mu + 3\lambda) \frac{dV}{dx} = 0. \qquad (2.45)$$

We expand now $V - V_1$ in powers of $p - p_1$ and $s - s_1$, taking pressure and entropy as independent variables.

$$
\begin{aligned}
V - V_1 &= \left(\frac{\partial V}{\partial p}\right)_s (p - p_1) + \frac{1}{2} \left(\frac{\partial^2 V}{\partial p^2}\right)_s (p - p_1)^2 \\
&+ \left(\frac{\partial V}{\partial s}\right)_p (s - s_1) + \frac{1}{2} \left(\frac{\partial^2 V}{\partial s^2}\right)_p (s - s_1)^2 .
\end{aligned}
$$

The coefficients are evaluated outside the shock wave, namely, for $p = p_1$, $s = s_1$. We will see that the entropy variation between a point upstream the shock and point inside the layer is at most of order of $(p_2 - p_1)^2$. The last term in the above equation can be neglected if we keep terms up to second order. By replacing the above expansion in Eq. 2.45, we have:

$$
\left[1 + j^2 \left(\frac{\partial V}{\partial p}\right)_s\right] (p - p_1) + \frac{1}{2} j^2 \left(\frac{\partial^2 V}{\partial p^2}\right)_s (p - p_1)^2 + j^2 \left(\frac{\partial V}{\partial s}\right)_p (s - s_1)
$$
$$
= j (2\mu + 3\lambda) \frac{dV}{dx}. \qquad (2.46)
$$

The derivative dV/dx can be written as:

$$\frac{dV}{dx} = \left(\frac{\partial V}{\partial p}\right)_s \frac{dp}{dx} + \left(\frac{\partial V}{\partial s}\right)_p \frac{ds}{dx}.$$

Let $p - p_1$ be a small number, of order $\mathcal{O}(1/\delta)$, where δ is the typical length, characterizing the thickness of the shock wave. Likewise, $V - V_1$ is of order $\mathcal{O}(1/\delta)$. Since:

$$V - V1 \approx \frac{dV}{dx} \delta = \mathcal{O}\left(\frac{1}{\delta}\right).$$

dV/dx must be of order $\mathcal{O}(1/\delta)$. Differentiation with respect to x increases thus, the order of the term. The derivative dp/dx is of second order and ds/dx of third order. The term can be neglected. Equation 2.46 becomes:

$$
\left[1 + j^2 \left(\frac{\partial V}{\partial p}\right)_s\right] (p - p_1) + \frac{1}{2} j^2 \left(\frac{\partial^2 V}{\partial p^2}\right)_s (p - p_1)^2 + j^2 \left(\frac{\partial V}{\partial s}\right)_p (s - s_1)
$$
$$
= -j (2\mu + 3\lambda) \left(\frac{\partial V}{\partial p}\right)_s \frac{dp}{dx}. \qquad (2.47)
$$

The energy equation is obtained from Eq. 5.15, dropping the time derivative, the gravitational term, internal heat sources, and expressing the heat flux in terms of Fourier's law $q = -\kappa \, dT/dx$. Integrating across the cross section and defining $u = Q/A$, where Q is the volumetric flow rate and A the area of the cross section, we obtain:

$$\frac{d}{dx}\left[\rho\left(e+\frac{u^2}{2}\right)u\right] = -\frac{d}{dx}\,(pu) + \frac{d}{dx}\,(\tau_{xx}u) + \frac{d}{dx}\kappa\frac{dT}{dx},$$

where e is the specific internal energy. Dropping the derivative operator d/dx, incorporating the pressure term to left-hand side term of the above equation, and replacing τ_{xx} by the expression given by the constitutive equation, we have:

$$\rho u\left(h+\frac{u^2}{2}\right) - (2\mu + 3\lambda)\,u - \kappa\frac{dT}{dx} = \text{Constant.}$$

Again, using the definition $u = jV$, we rewrite the energy equation in the form:

$$h + \frac{1}{2}j^2V^2 - j\,(2\mu + 3\lambda)\,V\frac{dV}{dx} - \frac{\kappa}{j}\frac{dT}{dx} = h_1 + \frac{1}{2}j^2V_1^2.$$

Rewriting the above equation:

$$(h-h_1) + \frac{1}{2}j^2\left(V^2 - V_1^2\right) - j\,(2\mu + 3\lambda)\,V\frac{dV}{dx} - \frac{\kappa}{j}\frac{dT}{dx} = h_1 + \frac{1}{2}j^2V_1^2. \tag{2.48}$$

We multiply each term of Eq. 2.45 by $(V_1 + V)/2$ and subtract the result from Eq. 2.48 to obtain:

$$(h - h_1) + \frac{1}{2}\,(V_1 + V)\,(p - p_1) - \frac{1}{2}j\,(2\mu + 3\lambda)\,(V + V_1)\frac{dV}{dx} - \frac{\kappa}{j}\frac{dT}{dx} = 0.$$

The viscous term in the above equation is of third order and can be neglected. We have then:

$$(h_1 - h_2) + \frac{1}{2}\,(V_1 + V)\,(p - p_1) - \frac{\kappa}{j}\frac{dT}{dx} = 0.$$

The terms $(h_1 - h_2) + (1/2)\,(V_1 + V)\,(p - p_1)$ can be obtained, up to second order, by Eq. 2.41, since first and second order in this expansion are zero. The terms $(h_1 - h_2) + (1/2)\,(V_1 + V)\,(p - p_1)$ are thus replaced by $T\,(s_2 - s_1)$. The derivative dT/dx is given by:

$$\frac{dT}{dx} = \left(\frac{\partial T}{\partial p}\right)_s\frac{dp}{dx} + \left(\frac{\partial T}{\partial s}\right)_p\frac{ds}{dx} \approx \left(\frac{\partial T}{\partial p}\right)_s\frac{dp}{dx},$$

resulting in:

$$T (s - s_1) = \frac{\kappa}{j} \frac{dT}{dx}. \tag{2.49}$$

Substituting the expression of $s - s_1$ from above in Eq. 2.46, we have:

$$\frac{1}{2} j^2 \left(\frac{\partial^2 V}{\partial p^2} \right)_s (p - p_1)^2 + \left[1 + j^2 \left(\frac{\partial V}{\partial p} \right)_s \right] (p - p_1)$$

$$= j \left[-\frac{\kappa}{T} \left(\frac{\partial V}{\partial s} \right)_p \left(\frac{\partial T}{\partial p} \right)_s + (2\mu + 3\lambda) \left(\frac{\partial V}{\partial p} \right)_s \right] \frac{dp}{dx}.$$

The flux j can be approximated by $j = u/V \approx a/V$, where a is the sound velocity. At great distances of the shock, the right-hand side of the above equation vanishes, since $dp/dx = 0$. The pressure is either p_1 or p_2. That said, we conclude that p_1 and p_2 are zeroes of the quadratic in p polynomial of the left-hand side. The polynomial is of the form $A (p - p_1)^2 + B (p - p_1)$, and can be factorized as the product $A (p - p_1) (p - p_2)$, leading to a differential equation for the function $p(x)$:

$$\frac{1}{2} \left(\frac{\partial^2 V}{\partial p^2} \right)_s (p - p_1) (p - p_2)$$

$$= -\frac{V^3}{a^3} \left[\frac{\kappa}{T} \left(\frac{\partial V}{\partial s} \right)_p \left(\frac{\partial T}{\partial p} \right)_s a^2 \rho^2 + (2\mu + 3\lambda) \right] \frac{dp}{dx}.$$

The term inside braces in the right-hand side of the above equation can be further simplified by noting that $(\partial V/\partial s)_p = (\partial T/\partial p)_s$. It can be shown that the coefficient of the term $-dp/dx$ is $2V^2 \beta$ [5], where

$$\beta = -\frac{V}{2a^3} \left[\frac{\kappa}{T} \left(\frac{\partial V}{\partial s} \right)_p \left(\frac{\partial T}{\partial p} \right)_s a^2 \rho^2 + (2\mu + 3\lambda) \right]. \tag{2.50}$$

It can also be shown that β is related to the sound absorption coefficient ξ and the sound frequency ω through the relation $\xi = \omega^2 \beta$ [5], with ξ defined by:

$$\xi = \frac{|\dot{E}_{mech}(s)|}{2a\bar{E}}, \tag{2.51}$$

where $\dot{E}_{mech}(s)$ is the rate of loss of the available energy that can be converted to work for the actual entropy s of the medium, a is the sound velocity, and $\bar{E} = \rho v_0^2$ is the energy content of the medium associated with a plane sound wave propagating in the x direction, by $v_x = v_0 \cos(\kappa x = \omega t)$, $v_y, v_z = 0$.

We have thus:

$$\frac{dp}{dx} = -\frac{1}{4V^2\beta} \left(\frac{\partial^2 V}{\partial p^2}\right)_s (p - p_1)(p - p_2)$$

Integration of this equation results in:

$$x = -\frac{4V^2\beta}{\left(\partial^2 V/\partial p^2\right)_s} \int \frac{dp}{(p - p_1)(p - p_2)} + \text{Constant}$$

$$= \frac{4V^2\beta}{(p_2 - p_1)\left(\partial^2 V/\partial p^2\right)_s /2} \tanh^{-1} \frac{p - (p_2 + p_1)/2}{\left(\partial^2 V/\partial p^2\right)_s /2} + \text{Constant}.$$

Setting the constant to zero, we find:

$$p - \frac{1}{2}(p_2 + p_1) = \frac{1}{2}(p_2 - p_1)\tanh\frac{x}{\delta}, \tag{2.52}$$

where:

$$\delta = 8\beta V^2 (p_2 - p_1) \left(\frac{\partial^2 V}{\partial p^2}\right)_s. \tag{2.53}$$

The jump across the shock is thus described by a hyperbolic tangent profile, with asymptotic values p_1 and p_2, as $c \to \pm\infty$, respectively. The point $x = 0$ corresponds to the median value of the pressure, $(p_1 + p_2)/2$. Most of the pressure variation occurs in a length of order of $\mathcal{O}(\delta)$, which can thus be called as the thickness of the shock wave. The thickness varies with the inverse of $p_1 + p_2$.

Entropy variations are directly obtained from Eqs. 2.52 and 2.49 [5]:

$$s - s_1 = \frac{\kappa}{16\alpha\beta V T} \left(\frac{\partial T}{\partial p}\right)_s \left(\frac{\partial^2 V}{\partial p^2}\right)_s (p_2 - p_1)^2 \frac{1}{\cosh(x/\delta)}. \tag{2.54}$$

This equation shows that the entropy does not vary monotonically, having a maximum at $x = o$. The formula also gives $s = s_1$ at $x \to \pm\infty$, what actually occurs at first order in $p_2 - p_1$, since entropy variations are of third order in Δp.

4   Rayleigh Line

This section addresses steady one-dimensional compressible flows of perfect gases in ducts with constant cross section, heat input or removal, and neglecting viscous effects between the flow and the duct walls. Heat input results in opposite effects on the flow velocity, depending on weather the flow is initially subsonic or supersonic.

The quantity $p + \rho u^2$ is conserved whereas the quantity $C_p + u^2/2$ changes due to the addition or removal of heat. For this reason, we denote $C_p + u^2/2$ as *total enthalpy*, and not by stagnation enthalpy. Similarly, the quantity $p + \rho u^2/2$ is no longer conserved, and is denoted by *total pressure*. The thermodynamic process undergone by the flow when heated or cooled is represented by a curve in the (h, s) plane. The curve is parameterized by the Mach number and denoted as a *Rayleigh Line* [1, 6, 7].

Steady one-dimensional inviscid gas flows in ducts with constant transversal section obey the momentum and the continuity equations in the form:

$$p_1 + \rho_1 u_1^2 = p_2 + \rho_1 u_2^2 \tag{2.55}$$

$$\rho_1 u_1 = \rho_2 u_2. \tag{2.56}$$

Adding an amount of heat q per unit of mass of the gas results in an increase of the total temperature, from T_{01} to T_{02}, given by:

$$q = C_p (T_{02} - T_{01}).$$

We derive the relations between the flow properties in two points of the duct, where the Mach numbers are, respectively, M_1 and M_2. The expression of ρu^2 can be written as:

$$\rho u^2 = \rho a^2 M^2 = \rho \frac{\gamma p}{\rho} M^2 = \gamma p M^2.$$

Introducing the above result in Eq. 2.55 we have:

$$p_2 - p_1 = \rho_1 u_1^2 - \rho_2 u_2^2 = \gamma p_1 M_1^2 - \gamma p_2 M_2^2,$$

which results in:

$$\frac{p_2}{p_1} = \frac{1 + \gamma M_1^2}{1 + \gamma M_2^2}. \tag{2.57}$$

For a perfect gas, and using Eq. 2.56, we have:

$$\frac{T_2}{T_1} = \frac{p_2 \, \rho_1}{p_1 \, \rho_2} = \frac{p_2 \, u_2}{p_1 \, u_1} = \frac{p_2}{p_1} \frac{M_2 a_2}{M_1 a_1} = \frac{p_2}{p_1} \frac{M_2}{M_1} \left(\frac{T_2}{T_1} \right)^{1/2}.$$

Introducing the expression of p_2/p_1, given by Eq. 2.57, in the above equation, we find:

$$\frac{T_2}{T_1} = \left(\frac{1 + \gamma M_1^2}{1 + \gamma M_2^2} \right)^2 \left(\frac{M_2}{M_1} \right)^2. \tag{2.58}$$

Since $\rho_2/\rho_1 = (p_2/p_1)\,(T_1/T_2)$, we obtain, using Eqs. 2.57 and 2.58:

$$\frac{\rho_2}{\rho_1} = \frac{1 + \gamma M_2^2}{1 + \gamma M_1^2} \left(\frac{M_1}{M_2}\right)^2 . \tag{2.59}$$

The total pressure ratio is obtained with Eqs. 2.10 and 2.57:

$$\frac{p_{02}}{p_{01}} = \frac{1 + \gamma M_1^2}{1 + \gamma M_2^2} \left(\frac{1 + \dfrac{\gamma - 1}{2} M_2^2}{1 + \dfrac{\gamma - 1}{2} M_1^2}\right)^{\gamma/(\gamma-1)} . \tag{2.60}$$

The ratio between total temperatures is obtained using Eqs. 2.8 and 2.57:

$$\frac{T_{02}}{T_{01}} = \frac{1 + \dfrac{\gamma - 1}{2} M_2^2}{1 + \dfrac{\gamma - 1}{2} M_1^2} \left(\frac{1 + \gamma M_1^2}{1 + \gamma M_2^2}\right)^2 \left(\frac{M_2}{M_1}\right)^2 . \tag{2.61}$$

For convenience, we adopt $M_1 = 1$ and represent $p_1 = P^*$, $\rho_1 = \rho^*$ $T_1 = T^*$, etc. Using the critical state as the reference state, we find the following ratios between properties at a state where the Mach number is M, and the critical ones:

$$\frac{p}{p^*} = \frac{1 + \gamma}{1 + \gamma M^2} \tag{2.62}$$

$$\frac{T}{T^*} = \left(\frac{1 + \gamma}{1 + \gamma M^2}\right)^2 M^2 \tag{2.63}$$

$$\frac{\rho}{\rho^*} = \frac{1 + \gamma M^2}{1 + \gamma} \frac{1}{M^2} \tag{2.64}$$

$$\frac{p_0}{p^*} = \frac{1 + \gamma}{1 + \gamma M^2} \left[\frac{2 + (\gamma - 1) M_2^2}{1 + \gamma}\right]^{\gamma/(\gamma-1)} \tag{2.65}$$

$$\frac{T_0}{T^*} = \frac{(1 + \gamma) M^2}{1 + \gamma M^2} \left[2 + (\gamma - 1) M^2\right]. \tag{2.66}$$

The behavior of the gas with heat addition is shown in the diagram of Fig. 2.5, which shows a plot of $(s - s^*)/R \times T/T^*$. The curve is parameterized with the Mach number associated with pairs of values of the above variables, lying on the curve. The curve presents two branches, the upper one being subsonic and the lower one supersonic. The flow entropy attains a maximum at $M = 1$. Since the entropy monotonically increases with the heat addition, a limit for the heat supplied

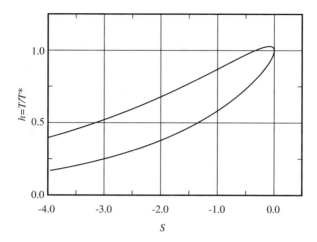

Fig. 2.5 Rayleigh line. The vertical axis contains the normalized specific enthalpy, $h = T/T^*$, and the horizontal one the normalized specific entropy, defined as $(s - s^*)/R$, where s and s^* are, respectively, the specific entropy at a given point of the curve, the one at the point where $M = 1$. We adopt $s^* = 0$ as the reference value. R is the perfect gases constant of the air. The flow is subsonic at the upper branch of the line, and supersonic at the lower one. Heat addition at the subsonic branch increases the flow Mach number until reaching $M = 1$, and decelerates in the supersonic regime. Further heating beyond the critical condition results in changes in the flow condition at the inlet of the constant transversal section duct

to the flow exists. At this maximum the Mach number of the flow is $M = 1$. Some qualitative features of steady one-dimensional compressible flows in ducts with constant transversal section and heat addition or removal are as follows [1].

In supersonic regime with heat addition:

1. The Mach number diminishes;
2. The pressure increases;
3. The temperature increases;
4. The total temperature increases;
5. The total pressure diminishes;
6. The velocity diminishes.

In subsonic regime with heat addition

1. The Mach number increases;
2. The pressure diminishes;
3. The temperature increases for $M < \gamma^{-1/2}$ and diminishes for $M > \gamma^{-1/2}$;
4. The total temperature increases;
5. The total pressure diminishes;
6. The velocity increases.

If heat is removed for the flow, the above effects are reversed. Progressive heat addition thus leads the flow to a limit condition where $M = 1$. The flow is said to be *chocked*. No further heat can be supplied to the flow without drastic changes in the upstream conditions. If flow is supersonic, with the regime attained through a convergent-divergent nozzle, supersonic regime, a shock wave is established at the divergent section of the nozzle, bringing the flow back to the subsonic regime. If the flow is initially subsonic, heat addition results in the propagation of upstream pressure waves and in the reduction of the Mach number until $M = 1$ is attained with the new total heat input.

It should be noted that it is possible to decelerate a gas flowing in a constant transversal section duct, from the supersonic to the subsonic regime, by first adding heat up to the critical condition where $M = 1$, and by the subsequent removal of heat. The inverse procedure allows the acceleration of a gas from the subsonic to the supersonic regime.

Finally, we point to the fact that heat addition, either in the subsonic or in the supersonic regime, always leads to reduction in the gas total pressure.

5 Fanno Line

This section addresses steady one-dimensional compressible flows of perfect gases in adiabatic ducts with constant cross section, resulting in constant total enthalpy, but considering momentum loss $\Delta \left(p + \rho u^2\right) < 0$. Following the same procedure adopted in Sect. 4, the thermodynamic process is represented by a curve in the (h, s) plane, parameterized by the Mach number. The curve is denoted by *Fanno Line* [1, 6, 7].

The one-dimensional flow of a perfect gas subjected to viscous forces at the walls of a constant transversal section duct is characterized by the Mach number and by two thermodynamic properties, suitably chosen. Usually, enthalpy h and entropy s are the chosen ones. Let D be duct diameter. The curve describing the gas thermodynamic state in the (h, s) plane parameterized by the Mach number is called as a *Fanno Line*. We assume that the viscous force between the one-dimensional flow and the duct walls is characterized by a constant Fanning factor f.

We derive now an equation relating the Mach number at two duct sections, located at a distance L. Both the mass and the energy, represented by the total enthalpy, are conserved across the elementary transversal section considered. Let dx be the length of the transversal section of the flow, as represented in Fig. 2.6, and F_A, the viscous force applied by the duct walls on the flow. The net balance of momentum, pressure, and viscous forces results in the following one-dimensional equation (Eq. 5.23):

$$dp + \rho u \, du + 4f \frac{dx}{D_H} \frac{\rho u^2}{2} = 0, \qquad (2.67)$$

Fig. 2.6 Balance of
Momentum fluxes and forces
in an elementary section of a
perfect gas flowing in a
constant transversal area duct

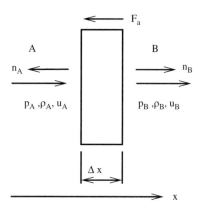

where D_H is the duct hydraulic diameter. The total energy equation becomes an equation of total enthalpy conservation. The equation reads:

$$C_p T_0 = C_p T + \frac{u^2}{2}.$$

In differential form:

$$C_p dT + \frac{1}{2} du^2 = 0.$$

By replacing $C_p = \gamma R/(\gamma - 1)$ we find:

$$\frac{\gamma R}{\gamma - 1} dT + \frac{1}{2} du^2 = \frac{a^2}{\gamma - 1} \frac{dT}{T} + \frac{1}{2} du^2 = 0,$$

what leads to

$$\frac{dT}{T} + \frac{\gamma - 1}{2} M^2 \frac{du^2}{u^2} = 0. \tag{2.68}$$

Using the definition of the Mach number, we write:

$$u^2 = a^2 M^2.$$

In differential form:

$$du^2 = a^2 dM^2 + M^2 da^2.$$

Dividing the last equation by u^2, we have:

$$\frac{du^2}{u^2} = \frac{dM^2}{M^2} + \frac{da^2}{a^2} = \frac{dM^2}{M^2} + \frac{dT}{T}. \tag{2.69}$$

The continuity equation can be written as:

$$\frac{d\rho}{\rho} + \frac{du}{u} = 0$$

or either:

$$\frac{d\rho}{\rho} + \frac{du^2}{2u^2} = 0. \tag{2.70}$$

For a perfect gas we have:

$$dp = \rho R dT + RT d\rho,$$

and also:

$$\frac{dp}{p} = \frac{dT}{T} + \frac{d\rho}{\rho}. \tag{2.71}$$

From Eqs. 2.68 and 2.71 we have:

$$\frac{dp}{p} - \frac{d\rho}{\rho} + \frac{\gamma - 1}{2} M^2 \frac{du^2}{u^2} = 0. \tag{2.72}$$

Using Eq. 2.70 we replace the term $d\rho/\rho$ of Eq. 2.72, and find:

$$\frac{dp}{p} + \frac{du^2}{2u^2} + \frac{\gamma - 1}{2} M^2 \frac{du^2}{u^2} = 0$$

or

$$\frac{dp}{p} + \left[\frac{1 + (\gamma - 1) M^2}{2} \right] \frac{du^2}{u^2} = 0. \tag{2.73}$$

The term $\left(\rho u^2 \right)/2$ can be written as:

$$\frac{1}{2}\rho u^2 = \frac{1}{2}\rho RT M^2 = \frac{\gamma}{2} p M^2.$$

Introducing this last result in Eq. 2.67 we find:

$$\frac{2}{\gamma M^2} \frac{dp}{p} + \frac{du^2}{2} = -\frac{4f}{D} dx.$$

We now replace the term dp/p of the above equation by the expression obtained from Eq. 2.72, to obtain:

$$-\frac{2}{\gamma M^2}\left[\frac{1 + (\gamma - 1)\, M^2}{2}\right]\frac{du^2}{u^2} + \frac{du^2}{2} = -\frac{4f}{D}dx.$$

Grouping terms:

$$\left[1 - \frac{1 + (\gamma - 1)\, M^2}{\gamma M^2}\right]\frac{du^2}{u^2} = -\frac{4f}{D}dx$$

and:

$$\frac{M^2 - 1}{\gamma M^2}\frac{du^2}{u^2} = -\frac{4f}{D}dx. \tag{2.74}$$

Introducing the expression of dT/T, obtained from Eq. 2.68, in Eq. 2.69 we have:

$$-\frac{\gamma - 1}{2}M^2\frac{du^2}{u^2} + \frac{dM^2}{M^2} = \frac{du^2}{u^2},$$

from where we obtain an equation for du^2/u^2:

$$\frac{du^2}{u^2} = \left(1 + \frac{\gamma - 1}{2}M^2\right)^{-1}\frac{dM^2}{M^2}.$$

We introduce now the expression of du^2/u^2 given by the above equation in Eq. 2.74 to obtain, after integration, an equation relating the Mach number at two sections of the duct, apart from a distance of length L:

$$\frac{1 - M^2}{\gamma M^2}\left(1 + \frac{\gamma - 1}{2}M^2\right)^{-1}\frac{dM^2}{M^2} = \frac{4f}{D}dx. \tag{2.75}$$

In terms of dM/M:

$$2\frac{1 - M^2}{\gamma M^2}\left(1 + \frac{\gamma - 1}{2}M^2\right)^{-1}\frac{dM}{M} = \frac{4f}{D}dx. \tag{2.76}$$

Proceeding with integration of Eq. 2.75:

$$\int_{M_1^2}^{M_2^2}\frac{1 - M^2}{\gamma M^4\left[1 + M^2\left(\gamma - 1\right)/2\right]}dM^2 = 4f\frac{L}{D}. \tag{2.77}$$

We now sum and subtract an expression to the numerator of the above equation to obtain:

$$1 - M^2 = 1 - M^2 + \left(1 - \frac{\gamma - 1}{2} M^2\right) - \left(1 - \frac{\gamma - 1}{2} M^2\right)$$

$$= 1 + \frac{\gamma - 1}{2} M^2 - \frac{\gamma + 1}{2} M^2.$$

Replacing the expression of $1 - M^2$ as above, in the integrand of Eq. 2.77, we have:

$$\int_{M_1^2}^{M_2^2} \frac{1 - M^2}{\gamma M^4 \left[1 + M^2 (\gamma - 1)/2\right]} dM^2$$

$$= \int_{M_1^2}^{M_2^2} \frac{dM^2}{\gamma M^4} - \int_{M_1^2}^{M_2^2} \frac{\gamma + 1}{2\gamma M^2 \left[1 + M^2 (\gamma - 1)/2\right]} dM^2. \qquad (2.78)$$

Upon defining:

$$y = \frac{M^2}{1 + M^2 (\gamma - 1)/2}$$

we have:

$$dy = \frac{dM^2}{1 + M^2 (\gamma - 1)/2} - \frac{1 + M^2 (\gamma - 1)/2}{\left[1 + M^2 (\gamma - 1)/2\right]^2} dM^2$$

$$= \frac{dM^2}{\left[1 + M^2 (\gamma - 1)/2\right]^2},$$

and

$$\frac{dy}{y} = \frac{dM^2}{M^2 \left[1 + M^2 (\gamma - 1)/2\right]}. \qquad (2.79)$$

By using the expression of dy/y, given by Eq. 2.79, and the decomposition given by Eq. 2.78, of the integral term of Eq. 2.77, we obtain successively:

$$4f \frac{L}{D} = \frac{1}{\gamma} \int_{M_1^2}^{M_2^2} \frac{dM^2}{M^2} - \frac{\gamma + 1}{2\gamma} \int_{y_1}^{y_2} \frac{dy}{y}$$

$$= \frac{1}{\gamma} \left(\frac{1}{M_1^2} - \frac{1}{M_2^2}\right) - \frac{\gamma + 1}{2\gamma} \ln \frac{y_1}{y_2},$$

and

$$4f\frac{L}{D} = \frac{1}{\gamma}\left(\frac{1}{M_1^2} - \frac{1}{M_2^2}\right) + \frac{\gamma+1}{2\gamma}\ln\left[\frac{M_1^2}{M_2^2}\frac{1+M_2^2\,(\gamma-1)/2}{1+M_1^2\,(\gamma-1)/2}\right].$$

The critical length of the duct, L^*, at the end of which $M = 1$ is attained, starting from a Mach number M, is given by:

$$4f\frac{L^*}{D} = \frac{1-M^2}{\gamma M^2} + \frac{\gamma+1}{2\gamma}\ln\frac{(\gamma+1)\,M^2}{2+(\gamma-1)\,M^2}. \tag{2.80}$$

We derive next the equations for the ratio between properties at two points of the duct, where the Mach number is, respectively, M_1 and M_2. The temperature ratio is obtained from Eq. 2.8:

$$\frac{T_2}{T_1} = \frac{T_2}{T_0}\frac{T_0}{T_1} = \frac{2+(\gamma-1)\,M_1^2}{2+(\gamma-1)\,M_2^2}. \tag{2.81}$$

The pressure ratio is obtained from the continuity equation $\rho_1 u_1 = \rho_2 u_2$ and the definition of the Mach number. We obtain:

$$\frac{\gamma p_1}{a_1^2} = \frac{\gamma p_2}{a_2^2}.$$

The pressure ratio is obtained from this last equation:

$$\frac{p_2}{p_1} = \frac{M_1\,a_2}{M_2\,a_1} = \frac{M_1}{M_2}\left(\frac{T_2}{T_1}\right)^{1/2}.$$

Using the expression of the temperature ratio at two points, given by Eq. 2.81, we find:

$$\frac{p_2}{p_1} = \frac{M_1}{M_2}\left[\frac{2+(\gamma-1)\,M_1^2}{2+(\gamma-1)\,M_2^2}\right]^{1/2}. \tag{2.82}$$

The ratio between specific masses is obtained by noting that:

$$\frac{\rho_2}{\rho_1} = \frac{p_2}{p_1}\frac{T_1}{T_2}.$$

Upon introducing the expression of the pressure and temperature ratios, we find:

$$\frac{\rho_2}{\rho_1} = \frac{M_1}{M_2}\left[\frac{2+(\gamma-1)\,M_2^2}{2+(\gamma-1)\,M_1^2}\right]^{1/2}. \tag{2.83}$$

The ratio between the total pressures is obtained from Eqs. 2.10 and 2.82. We obtain:

$$\frac{p_{02}}{p_{01}} = \frac{M_1}{M_2} \left[\frac{2 + (\gamma - 1) M_2^2}{2 + (\gamma - 1) M_1^2} \right]^{(\gamma+1)/2(\gamma-1)} . \tag{2.84}$$

The ratio between the properties at a prescribed point of the duct and the point where $M = 1$ are given below:

$$\frac{T}{T^*} = \frac{\gamma + 1}{2 + (\gamma - 1) M^2} \tag{2.85}$$

$$\frac{p}{p^*} = \frac{1}{M} \left[\frac{\gamma + 1}{2 + (\gamma - 1) M^2} \right]^{1/2} \tag{2.86}$$

$$\frac{\rho}{\rho^*} = \frac{1}{M} \left[\frac{2 + (\gamma - 1) M^2}{\gamma + 1} \right]^{1/2} \tag{2.87}$$

$$\frac{p_0}{p_0^*} = \frac{1}{M} \left[\frac{2 + (\gamma - 1) M^2}{\gamma + 1} \right]^{(\gamma+1)/2(\gamma-1)} . \tag{2.88}$$

The Rayleigh and Fanno lines are shown in Fig. 2.7. The vertical axis contains the same variables used in Fig. 2.5. In both curves, the upper branch corresponds to subsonic and the lower ones to supersonic flow. The intersection of the two curves satisfies the conditions for the existence of normal shocks. The shock occurs with a transition from the supersonic to the subsonic regime, in order to comply with the restriction imposed by the second law of thermodynamics.

Some distinguished characteristics of one-dimensional compressible flows subjected to viscous effects are [1]:

In the supersonic regime:

1. The Mach number diminishes;
2. The pressure increases;
3. The temperature increases;
4. The total pressure diminishes;
5. The velocity diminishes.

In the subsonic regime:

1. The Mach number increases;
2. The pressure diminishes;
3. The temperature diminishes;
4. The total pressure diminishes;
5. The velocity increases.

The viscous friction as well as heat addition decelerate initially the supersonic flow and accelerates initially the subsonic flows up to the critical point where $M = 1$ is attained, after covering a distance L^*, given by Eq. 2.80. The addition of an

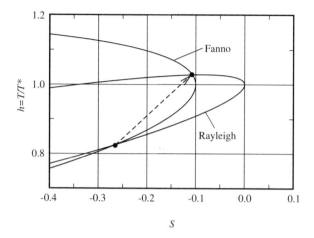

Fig. 2.7 Rayleigh and Fanno Lines. The vertical axis contains the normalized specific enthalpy, $h = T/T^*$, and the horizontal one the normalized specific entropy, defined as $(s - s^*)/R$, where s and s^* are, respectively, the flow specific entropy at a point of the curve, and at the point where $M = 1$. We adopt $s^* = 0$ as the reference value for the Rayleigh Line, and $s^* = -0.1$ for the Fanno Line. R is the perfect gas constant of the air. The upper branch of each curve corresponds to the subsonic regime and the lower one to the supersonic. Viscous friction of the one-dimensional flow with the duct walls as well as heat addition decelerate initially the supersonic flows and accelerate the subsonic ones, up to the point where $M = 1$. The intersection points of the two curves satisfy the conditions for the existence of a normal shock. The shock occurs with a transition from the supersonic to the subsonic regime, in order to comply with the restriction imposed by the second law of thermodynamics

additional duct length after the critical point with $M = 1$ is attained leads to similar phenomena observed with heat supply beyond the critical point: if the supersonic regime is attained through an expansion in a convergent-divergent nozzle, a shock wave establishes at the divergent section of the nozzle, bringing the flow to the subsonic regime. If the flow is initially subsonic, an increase in the duct length beyond the critical one leads to the reduction of the inlet Mach number and mass flow rate.

Supersonic flows in ducts with constant transversal section can be decelerated to $M = 1$ by adding heat, and further decelerated to the subsonic regime by removing heat. Similar deceleration from the supersonic to the subsonic regime cannot be made by using friction effects.

6   Friction in Ducts with Variable Cross Section

This section addresses the problem of compressible fluid flow in ducts with variable cross section, subjected to viscous friction between the walls and the flow. The problem is governed by the continuity and energy conservation, the last one represented by enthalpy conservation. Additionally, the applicable momentum equation is given by (see Eq. 5.24):

$$dp + \rho u \, du + 4f \frac{dx}{D_H} \frac{\rho u^2}{2} = 0,$$

which can be rewritten as:

$$\frac{dp}{p} + \gamma M^2 \frac{du}{u} + \frac{\gamma M^2}{2} 4f \frac{dx}{D_H} = 0. \tag{2.89}$$

From the continuity and the perfect gas equations, we can write:

$$\frac{dp}{p} = \frac{d\rho}{\rho} + \frac{dT}{T} = -\frac{dA}{A} - \frac{du}{u} + \frac{dT}{T}.$$

Introducing this result in Eq. 2.89, we find:

$$-\frac{dA}{A} - \frac{du}{u} + \frac{dT}{T} + \gamma M^2 \frac{du}{u} + \frac{\gamma M^2}{2} 4f \frac{dx}{D_H} = 0. \tag{2.90}$$

From the definition of Mach number, we have:

$$\frac{du}{u} = \frac{1}{2} \frac{dT}{T} + \frac{dM}{M}.$$

Substituting this result in Eq. 2.90, we obtain:

$$-\frac{dA}{A} - \frac{dM}{M} + \frac{1}{2} \frac{dT}{T} + \frac{\gamma M^2}{2} \frac{dT}{T} + \gamma M^2 \frac{dM}{M} + \frac{\gamma M^2}{2} 4f \frac{dx}{D_H} = 0.$$

Grouping terms:

$$-\frac{dA}{A} + (\gamma M^2 - 1) \frac{dM}{M} + \left(\frac{\gamma M^2 + 1}{2} \right) \frac{dT}{T} + \frac{\gamma M^2}{2} 4f \frac{dx}{D_H} = 0. \tag{2.91}$$

We make use now of the hypothesis that the energy of the flow, expressed by the total enthalpy, is conserved, namely, that $C_p T_0 = C_p T + u^2/2 = $ Constant. From this equation, we can write:

$$C_p dT + u \, du = 0,$$

which can be rewritten as:

$$\frac{dT}{T} + \frac{u \, du}{C_p T} = 0. \tag{2.92}$$

Since:

$$M^2 = \frac{u^2}{\gamma RT} \qquad \Longrightarrow \qquad 2M\,dM = \frac{2u\,du}{\gamma RT} - \frac{u^2}{\gamma RT^2}dT.$$

Rearranging terms, we have:

$$u\,du = \gamma RT\,M\,dM + \frac{u^2}{2}\frac{dT}{T}.$$

Introducing this last result in Eq. 2.92, we find:

$$\frac{dT}{T} + \frac{\gamma R}{C_p}M\,dM + \frac{u^2}{2C_pT}\frac{dT}{T} = 0. \tag{2.93}$$

Since $C_p - C_v = R$, we have $C_p = \gamma R/(\gamma - 1)$. Introducing the last expression for C_p in Eq. 2.93 we have:

$$\frac{dT}{T} + (\gamma - 1)\,M\,dM + \frac{\gamma - 1}{2}M^2\frac{dT}{T} = 0.$$

This equation can be rearranged to give:

$$\frac{dT}{T} = -\frac{(\gamma - 1)\,M^2}{1 + (\gamma - 1)\,M^2/2}\frac{dM}{M}. \tag{2.94}$$

Introducing the expression of dT/T, given by Eq. 2.94, into Eq. 2.91, and simplifying the terms we obtain:

$$-\frac{dA}{A} + \frac{M^2 - 1}{1 + (\gamma - 1)\,M^2/2}\frac{dM}{M} + \frac{\gamma M^2}{2}4f\frac{dx}{D_H} = 0.$$

This equation can be rewritten in the form:

$$\frac{dM}{M} = -\left[\frac{1 + (\gamma - 1)\,M^2/2}{1 - M^2}\right]\frac{dA}{A} + \left[\frac{1 + (\gamma - 1)\,M^2/2}{1 - M^2}\right]\frac{\gamma M^2}{2}4f\frac{dx}{D_H}. \tag{2.95}$$

The above equation shows that changes in the Mach number are due to area changes, as described by the first term on the right-hand side of this equation or to viscosity effects, described by the second term. When the area is constant, we recover the equations for the Fanno Line, and when friction is neglected the isentropic equations for flow in variable transversal area are recovered. We also point that the Mach number can be constant in ducts with variable area, provided that viscous effects just balance area changes. This condition is expressed by:

$$\frac{dA}{dx} = \frac{2f\gamma M^2 A}{D_H}.$$

According to this equation the area of the transverse must always increase, whenever friction with the walls is present, in order to keep the Mach number constant.

Equation 2.95 can be rewritten in the form:

$$\left(1 - M^2\right)\frac{dM}{M} = -\left(1 + \frac{\gamma - 1}{2}M^2\right)\frac{dA}{A} + \left(1 + \frac{\gamma - 1}{2}M^2\right)\frac{\gamma M^2}{2}4f\frac{dx}{D_H}.$$

(2.96)

This equation shows that in a point of the duct where the flow attains critical condition, with $M = 1$, the following equation holds:

$$-\left(1 + \frac{\gamma - 1}{2}M^2\right)\frac{dA}{A} + \left(1 + \frac{\gamma - 1}{2}M^2\right)\frac{\gamma M^2}{2}4f\frac{dx}{D_H} = 0.$$

If friction between the duct walls and the flow is negligible, $M = 1$ is attained in a duct throat, where $dA/A = 0$, as discussed in Sect. 2. However, the above equation shows that in flows subjected to friction forces at the walls, the critical condition occurs in duct points where $dA/A > 0$.

7 Heat Transfer in Ducts with Variable Cross Section

This section deals with the question of how a one-dimensional flow of a perfect gas evolves along a duct with variable cross section, when heat is added or removed to the fluid. Friction with the walls is neglected. The applicable continuity, momentum, and constitutive perfect gas equations can be written in the form:

$$\frac{d\rho}{\rho} + \frac{dA}{A} + \frac{du}{u} = 0$$

(2.97)

$$dp + \rho u\,du = 0$$

(2.98)

$$\frac{dp}{p} = \frac{d\rho}{\rho} + \frac{dT}{T},$$

(2.99)

whereas the energy conservation reads:

$$dq = C_p dT_0,$$

(2.100)

where dq stands for the heat supplied to the flow through the walls of an elementary transverse section of the duct, and:

$$dT_0 = dT + \frac{u\,du}{C_p}.$$

Introducing the expression of dp, obtained from Eq. 2.98 in Eq. 2.99, we find:

$$\frac{d\rho}{\rho} + \frac{dT}{T} + \frac{\rho u \, du}{p} = 0.$$

Since $a^2 = \gamma p/\rho$ we rewrite the last equation by:

$$\frac{d\rho}{\rho} + \frac{dT}{T} + \gamma M^2 \frac{du}{u} = 0.$$

We introduce now the expression of $d\rho/\rho$, given by Eq. 2.97 in the above equation, to obtain:

$$-\frac{dA}{A} + \frac{dT}{T} + \left(\gamma M^2 - 1\right)\frac{du}{u} = 0.$$

Taking into account that $u^2 = M^2 \gamma RT$, we can write, in differential form:

$$u \, du = \gamma RT M \, dM + \frac{M^2}{2}\gamma RT \, dT,$$

from which we have:

$$\frac{du}{u} = \frac{dM}{M} + \frac{1}{2}\frac{dT}{T}. \qquad (2.101)$$

We replace the term du/u in Eq. 7 by the expression obtained from the above one to obtain:

$$-\frac{dA}{A} + \frac{\gamma M^2 + 1}{2}\frac{dT}{T} + \left(\gamma M^2 - 1\right)\frac{dM}{M}. \qquad (2.102)$$

Using Eq. 2.8, we write:

$$T_0 = T\left(1 + \frac{\gamma - 1}{2}M^2\right). \qquad (2.103)$$

In differential form:

$$\frac{dT_0}{T} = \left(1 + \frac{\gamma - 1}{2}M^2\right)\frac{dT}{T} + (\gamma - 1)M^2\frac{dM}{M}.$$

Using Eq. 2.100, we rewrite the above equation as:

$$\frac{dq}{C_p T} = \left(1 + \frac{\gamma - 1}{2}M^2\right)\frac{dT}{T} + (\gamma - 1)M^2\frac{dM}{M}.$$

From this last equation, we obtain an expression for dT/T:

$$\frac{dT}{T} = \frac{1}{1 + (\gamma - 1) M^2/2} \frac{dq}{C_p T} - \frac{(\gamma - 1) M^2}{1 + (\gamma - 1) M^2/2} \frac{dM}{M}. \tag{2.104}$$

Replacing the expression of dT/T, obtained from the above equation, in Eq. 2.102, we have:

$$-\frac{dA}{A} + \frac{(\gamma M^2 + 1)/2}{1 + (\gamma - 1) M^2/2} \frac{dq}{T} + \frac{M^2 - 1}{1 + (\gamma - 1) M^2/2} \frac{dM}{M} = 0,$$

which can also be written as:

$$-\frac{dA}{A} + \frac{(\gamma M^2 + 1)/2}{1 + (\gamma - 1) M^2/2} \frac{dT_0}{T} + \frac{M^2 - 1}{1 + (\gamma - 1) M^2/2} \frac{dM}{M} = 0.$$

The first term in the above equation represents the effect of the transversal area change on the Mach number, whereas the second one copes with the influence of the heat transfer on the Mach number. If the flow is adiabatic, the critical condition corresponding to $M = 1$ occurs at the duct minimum area. However, if the flow is subjected to heat transfer, the critical condition occurs if:

$$-\frac{dA}{A} + \frac{(\gamma + 1)/2}{1 + (\gamma - 1)/2} \frac{dq}{T},$$

namely $M = 1$ occurs at a divergent point of the duct, where $dA/A > 0$, if heat is added to the flow, and at convergent point, where heat is removed.

8 Friction and Heat Transfer in Ducts with Constant Cross Section

This section addresses the problem of the flow of a perfect gas in ducts with constant cross section subjected to friction and heat transfer from the walls. The applicable continuity and momentum equations, the last one given by Eq. 5.24, are here rewritten:

$$\frac{d\rho}{\rho} + \frac{du}{u} = 0 \tag{2.105}$$

$$\frac{dp}{p} + \gamma M^2 \frac{du}{u} + \frac{\gamma M^2}{2} 4f \frac{dx}{D_H} = 0. \tag{2.106}$$

We replace the factor du/u by the expression given by Eq. 2.101 to obtain:

$$\frac{dp}{p} + \gamma M^2 \frac{dM}{M} + \frac{\gamma M^2}{2} \frac{dT}{T} + \frac{\gamma M^2}{2} 4f \frac{dx}{D_H} = 0. \tag{2.107}$$

The constitutive equation of perfect gases given by Eq. 2.99 is rewritten taking into account Eq. 2.105:

$$\frac{dp}{p} = -\frac{du}{u} + \frac{dT}{T}.$$

Replacing the expression of du/u, given by Eq. 2.101, we have:

$$\frac{dp}{p} = -\frac{dM}{M} + \frac{1}{2}\frac{dT}{T}.$$

The expression of dp/p, given in this last equation, is introduced in Eq. 2.107, leading to:

$$\frac{1+\gamma M^2}{2}\frac{dT}{T} + \frac{\gamma M^2}{2}4f\frac{dx}{D_H} - \left(1-\gamma M^2\right)\frac{dM}{M} = 0. \tag{2.108}$$

We substitute now the expression of dT/T, given by Eq. 2.104 in the above equation to find:

$$\frac{\left(1+\gamma M^2\right)/2}{1+(\gamma-1)M^2/2}\frac{dq}{C_pT} + \frac{\gamma M^2}{2}4f\frac{dx}{D_H}$$

$$= \left[\left(1-\gamma M^2\right) + \frac{1+\gamma M^2}{2}\frac{(\gamma-1)M^2}{1+(\gamma-1)M^2/2}\right]\frac{dM}{M}.$$

This equation can be further simplified by expressing the temperature T as a function of the local stagnation temperature (or local total temperature), given by Eq. 2.103:

$$\frac{1+\gamma M^2}{2}\frac{dq}{C_pT_0} + \frac{\gamma M^2}{2}$$

$$= \left[\left(1-\gamma M^2\right) + \frac{1+\gamma M^2}{2}\frac{(\gamma-1)M^2}{1+(\gamma-1)M^2/2}\right]\frac{dM}{M}. \tag{2.109}$$

The two terms on the left-hand side of the above equation represent, respectively, the contribution of heat transfer and of friction with the walls, for variations of the Mach number.

9   Friction in Isothermal Ducts with Constant Cross Section

A common case of compressible fluid flow occurs in the transport of gases through long pipelines, not heavily insulated, where the walls' temperature remains approximately constant. In these cases, heat transfer through the duct walls is usually small, whereas pressure drops due to friction with the walls cannot be neglected due to the usual long lengths of pipelines. The present section addresses this problem.

The flow momentum is governed by the following equation, Eq. 5.24:

$$\frac{dp}{p} + \gamma M^2 \frac{du}{u} + \frac{2f}{D_H}\gamma M^2 dx = 0. \tag{2.110}$$

Assuming constant area of the transverse section and temperature, we have, respectively, from the continuity, perfect gas law and the definition of the Mach number:

$$\frac{du}{u} + \frac{d\rho}{\rho} = 0$$

$$\frac{dp}{p} = \frac{d\rho}{\rho}$$

$$\frac{dM}{M} = \frac{du}{u}.$$

From these equations, we have:

$$\frac{du}{u} = -\frac{d\rho}{\rho} = -\frac{dp}{p} = \frac{dM}{M}. \tag{2.111}$$

Introduction of Eq. 2.110 in 2.111 leads to:

$$-\frac{dp}{p} = \frac{du}{u} = \left(\frac{\gamma M^2}{1 - \gamma M^2}\right)\frac{2f}{D_H}dx. \tag{2.112}$$

Using the result from Eq. 5.30 in the above one, we obtain:

$$\frac{dq}{u^2} = \frac{du}{u} = -\frac{dp}{p} = -\frac{d\rho}{\rho} = \frac{dM}{M} = \left(\frac{\gamma M^2}{1 - \gamma M^2}\right)\frac{2f}{D_H}dx. \tag{2.113}$$

From Eq. 5.31, we have:

$$\frac{ds}{C_p T} = \frac{dT}{T} - \frac{\gamma - 1}{\gamma}\frac{dp}{p} = \frac{\gamma - 1}{\gamma}\frac{dp}{p}. \tag{2.114}$$

Introducing the result from 2.112 into Eq. 2.114, we finally obtain:

$$\frac{ds}{C_p T} = \frac{\gamma - 1}{\gamma}\frac{dp}{p}\left(\frac{\gamma M^2}{1 - \gamma M^2}\right)\frac{2f}{D_H}dx. \tag{2.115}$$

Table 2.1 Effect of Mach number on the signal of changes of variables in isothermal flow of gases in constant area ducts

M	dq	dp	$d\rho$	du	dM	ds
$M < 1/\sqrt{\gamma}$	+	−	−	+	+	+
$M > 1/\sqrt{\gamma}$	−	+	+	−	−	−

Equations 2.113 and 2.115 have in common a term with the factor $1 - \gamma M^2$ in the denominator. Since all other factors in the right side of both equations are positive dp, du, etc. change sign when the Mach number surmount the value $1/\sqrt{\gamma}$. A summary of the signal of changes in the flow variables is given in Table 2.1.

As $M \to 1/\sqrt{\gamma}$ the amount of heat to be added or removed from the flow tends to infinite, to keep the gas temperature constant. Equation 2.113 shows that the Mach number increases in the direction of the flow if $M < 1/\sqrt{\gamma}$, and decreases otherwise. In consequence, $1/\sqrt{\gamma}$ is a limit value for the Mach number of gases flowing in isothermal ducts with constant transversal section.

Equation 2.111 can be integrated leading to:

$$\frac{u_2}{u_1} = \frac{\rho_2}{\rho_1} = \frac{p_2}{p_1} = \frac{M2}{M_1}. \tag{2.116}$$

From Eq. 2.113 we can write:

$$\frac{dM}{\gamma M^3} - \frac{DM}{M} = \frac{2f}{D_H}dx.$$

This equation can be integrated between points 1 and 2 of the duct, leading to:

$$\left(\frac{1}{2\gamma M_1^2} - \frac{1}{2\gamma M_2^2}\right) - \ln\frac{M_2}{M_1} = \frac{2fl}{D_H}. \tag{2.117}$$

The maximum length l^+ achievable with prescribed inlet Mach number M_1 is obtained for $M_2 = 1/\sqrt{\gamma}$. In this case, Eq. 2.117 becomes:

$$\left(\frac{1}{2\gamma M^2} - \frac{1}{2}\right) + \ln\sqrt{\gamma}M = \frac{2fl^+}{D_H}$$

or:

$$\left(\frac{1 - \gamma M^2}{\gamma M^2}\right) + \ln\left(\gamma M^2\right) = \frac{4fl^+}{D_H}, \tag{2.118}$$

where M stands for the inlet Mach number. No solution exists for $l > l^+$, with the prescribed inlet Mach number M. If $l < l^+$, the Mach number M_2 at $x = l$ is defined by Eq. 2.117 for given M_1. The ratio M_2/M_1 is determined and so are the ratios p_2/p_1, ρ_2/ρ_1 and u_2/u_1.

The critical ratio between these variables is obtained for $M_2 = 1/\sqrt{\gamma}$, and using Eq. 2.116:

$$\frac{u^+}{u} = \frac{\rho}{\rho^+} = \frac{p}{p^+} = \frac{M^+}{M} = \frac{1}{\sqrt{\gamma}M}. \qquad (2.119)$$

The quantities $4fl^+/D_H$, u/u^+, ρ/ρ^+, and p/p^+ depend on the inlet Mach number M, and on the adiabatic exponent γ only. The ratio between the inlet stagnation temperature T_0 and the stagnation temperature T_0^+, at the critical length l^+, where the Mach number is $M^+ = 1/\sqrt{\gamma}$ can be found using the isentropic relation:

$$\frac{T_0}{T} = 1 + \frac{\gamma - 1}{2}M^2.$$

The ratio T_0/T_0^+ is thus:

$$\frac{T_0}{T_0^+} = \frac{1 + (\gamma - 1)\,M^2/2}{1 + [(\gamma - 1)/2]\,(1/\gamma)}$$

or:

$$\frac{T_0}{T_0^+} = \frac{2\gamma}{3\gamma - 1}\left(1 + \frac{\gamma - 1}{2}M^2\right).$$

A plot of relevant variables of isothermal gas flow in ducts with constant area, friction, and heat transfer is shown in Fig. 2.8.

Fig. 2.8 Properties of isothermal gas flow in ducts with constant area, friction, and heat transfer for prescribed inlet Mach number M_1 and $\gamma = 1.4$ (see problem 6)

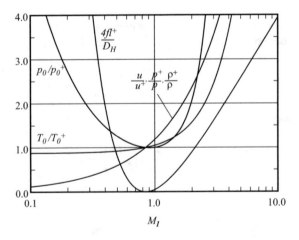

10 Analogy with Open Channel Hydraulics

10.1 The Area × Velocity Relation

This section discusses the analogy between steady one-dimensional compressible flow, addressed in Sects. 2 and 4, and the inviscid hydraulics in shallow channels with variable transversal section, as shown in Fig. 2.9.

The hypothesis of open channel flow with the upper surface in contact with air appears in the assumption of constant atmospheric pressure at the upper surface of the flow, independent of the thickness h of the fluid layer. Since we deal with quasi-one-dimensional flows, only the u component of the velocity, taken as the ratio $u = Q/A$, where Q is the volumetric flow rate and A, the cross section area, is considered. Within this assumption the momentum equation along the vertical direction reads:

$$dp = -\rho g \, dz.$$

Integration of this equation leads to the vertical distribution of pressures in the form:

$$p = \rho g (h - z).$$

We consider a control volume with length Δx, perpendicular to the flow direction and denote l and h as the channel width and the flow layer thickness, respectively. Application of the mass conservation principle to the control volume leads to $Q = C^{te}$ or $uhl = C^{te}$. This equation can also be written as:

$$\frac{du}{u} + \frac{dl}{l} + \frac{dh}{h} = 0$$

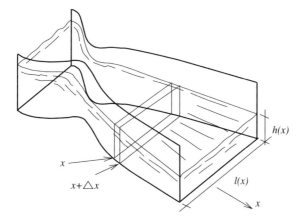

Fig. 2.9 Flow in a channel with variable cross section and free surface

or

$$\frac{dh}{h} = -\frac{du}{u} - \frac{dl}{l}. \tag{2.120}$$

Balance of flow momentum in the control volume can be expressed by:

$$\begin{pmatrix} \text{Rate of accumulation} \\ \text{of momentum in the} \\ \text{control volume} \end{pmatrix} = - \begin{pmatrix} \text{Net rate of loss of} \\ \text{momentum by the con-} \\ \text{trol volume} \end{pmatrix}$$
$$+ \begin{pmatrix} \text{Resultant of the actuat-} \\ \text{ing forces} \end{pmatrix}.$$

We now apply the momentum balance equation to the control volume. Since we consider the steady state the net balance of momentum transfer vanishes. The flow of momentum to the control volume is given by:

$$\int_0^h \rho u^2 l \, dz = \rho u^2 l h = f(x).$$

The rate of transfer of momentum leaving the control volume at $x + \Delta x$ reads:

$$-f(x + \Delta x) = -\left[f(x) + \frac{df}{dx} \Delta x \right] = -\left[\rho u^2 l h + \frac{d}{dx} (\rho u^2 l h) \Delta x \right].$$

The net flow of momentum to the control volume is given by the difference between the incoming minus the fluxes transferred to the control volume, and leaving the volume.

$$-\frac{d}{dx} (\rho u^2 l h) \Delta x.$$

The incompressibility condition is expressed by imposing uhl is constant. Introducing this condition in the equation of momentum balance we obtain:

$$-\frac{d}{dx} (\rho u^2 l h) \Delta x = -\rho u l h \frac{du}{dx} \Delta x.$$

The forces acting on the control volume are due to pressure at each side. The force $F(x)$ acting at the considerate section is given by the pressure integral at each height, $\rho g (h - z)$, multiplied by the area element $l \, dz$:

$$F(x) = \int_0^h \rho g (h - z) l \, dz = \rho g l \left(h z - \frac{z^2}{2} \right) \Big|_0^h = \rho g l \frac{h^2}{2}.$$

Forces actuating at $x + \Delta x$ is given by $F(x + \Delta x) = -[F(x) + dF/dx\,\Delta x]$, and the resultant is thus:

$$-\frac{d}{dx}F(x)\Delta x = -\rho g \frac{d}{dx}\left(l\frac{h^2}{2}\right)\Delta x.$$

The momentum balance becomes:

$$0 = -\rho u l h \frac{du}{dx}\Delta x - \rho g \frac{d}{dx}\left(l\frac{h^2}{2}\right)\Delta x$$

or

$$u l h \frac{du}{dx} = -g\frac{d}{dx}\left(l\frac{h^2}{2}\right) = -gl\frac{d}{dx}\frac{h^2}{2} - g\frac{h^2}{2}\frac{dl}{dx} = -glh\frac{dh}{dx} - g\frac{h^2}{2}\frac{dl}{dx}.$$

Multiplying this last equation by $dx/(glh^2)$, we have:

$$\frac{u}{gh}du = -\frac{dh}{h} - \frac{1}{2}\frac{dl}{l}.$$

Replacing dh/h by the expression given in Eq. 2.120 we obtain:

$$\frac{u}{gh}du = \frac{du}{u} + \frac{dl}{l} - \frac{1}{2}\frac{dl}{l}$$

or

$$\frac{u}{gh}du - \frac{du}{u} = \frac{1}{2}\frac{dl}{l}.$$

Putting the du/u of the left-hand side of the above equation in evidence we find:

$$\left(\frac{u^2}{gh} - 1\right)\frac{du}{u} = \frac{1}{2}\frac{dl}{l}.$$

The term u^2/gh is the square of the Froude number, Fr^2. Using this definition, we have:

$$\left(Fr^2 - 1\right)\frac{du}{u} = \frac{1}{2}\frac{dl}{l}. \tag{2.121}$$

Comparison between this equation and Eq. 2.4 evidences the analogy between the flow of liquids in open channels and compressible flows. The channel width l plays the role of the area A of the duct transverse section in compressible flows, whereas the depth h of the liquid layer is equivalent to the gas specific mass ρ. The Froude number is equivalent to the Mach number in compressible flows. For Froude numbers smaller than one (subcritical flows), enlargement of the channel results in the reduction of the liquid velocity and in the increase of the thickness of the fluid layer. For Froude numbers greater than one (supercritical flows), channel enlargements further increase the liquid velocity and reduces the thickness of the layer. Reversible critical flows with $Fr = 1$ can occur in sections where $dl/dx = 0$, in analogy to reversible compressible flows, which can only occur in sections where $dA/dx = 0$.

10.2   Shallow Water Theory: Hydraulic Jumps

As discussed in the previous section, flow of water in an open channel with variable transverse section is qualitatively similar to compressible gas flow through convergent and/or divergent nozzles. In this section we address the problem of *hydraulic jumps*, where a discontinuity in the height h of a water layer occurs, with a sudden increase in the height, and reduction of the flow velocity, from a value larger than the velocity $a = \sqrt{gh}$ of propagation of small perturbations in the free surface, to a smaller one. Hydraulic jumps are thus analogous to shock waves in compressible flows.

The momentum flow per unit of the channel width is given by:

$$\int_0^h \left(p + \rho u^2 \right) dz = \int_0^h \rho \left[g \left(h - z \right) + u^2 \right] dz = \frac{1}{2} \rho g h^2 + \rho u^2 h.$$

Since we assume that the hydraulic jump occurs at constant transversal width of the channel, mass and momentum conservation across the jump are expressed by:

$$u_1 h_1 = u_2 h_2 \tag{2.122}$$

$$\frac{1}{2} g h_1^2 + u_1^2 h_1 = \frac{1}{2} g h_2^2 + u_2^2 h_2, \tag{2.123}$$

where subscript 1 applies to quantities upstream and subscript 2 to quantities downstream the hydraulic jump. The above equations relate four variables, u_1, u_2, h_1, and h_2, two of which can be arbitrarily specified. Expressing the velocities v_1 and v_2 in terms of the heights h_1 and h_2, we find successively:

$$u_1^2 = \frac{1}{2} g \frac{h_2^2 - h_1^2}{h_1} + u_2^2 \frac{h_1}{h_2},$$

and:

$$u_1^2 \frac{h_2 - h_1}{h_1} = \frac{1}{2} g \frac{(h_2 + h_1)(h_2 - h_1)}{h_1}$$

and finally, the expressions for the velocities up, and analogously, downstream, u_1^2 and u_2^2:

$$u_1^2 = \frac{1}{2} g h_2 \frac{h_1 + h_2}{h_1} \tag{2.124}$$

$$u_2^2 = \frac{1}{2} g h_1 \frac{h_1 + h_2}{h_2}. \tag{2.125}$$

The mechanical energy fluxes on the two sides of the discontinuity are not the same, and the difference amounts to the energy dissipated in the discontinuity, per unit of time. The energy flux density in the channel is given by:

$$q = \int_0^h \left(\frac{p}{\rho} + \frac{1}{2} u^2 \right) \rho u \, dz = \int_0^h \left[g(h - z) + \frac{1}{2} u^2 \right] \rho u \, dz$$

$$= \frac{\rho u h}{2} \left(g h + u^2 \right) = \frac{j}{2} \left(g h + u^2 \right),$$

where $j = \rho u h$. The difference between the mechanical energy at the two sides of the jump is given by:

$$q_1 - q_2 = \frac{j}{2} \left[\left(g h_1 + u_1^2 \right) - \left(g h_2 + u_2^2 \right) \right].$$

Upon replacing u_1^2 and u_2^2 by the expressions given by Eqs. 2.124 and 2.125, we find:

$$q_1 - q_2 = g j \frac{(h_1^2 + h_2^2)(h_2 - h_1)}{4 h_1 h_2}.$$

Since part of the mechanical energy is dissipated across the jump $q_1 - q_2$ must be positive, implying in $h_2 > h_1$, namely, the fluid moves from the smaller to the higher height, across the jump. We can also see from Eqs. 2.124 and 2.125 that:

$$u_1 > c_1 = \sqrt{g h_1} \qquad \text{and} \qquad u_2 < c_2 = \sqrt{g h_2},$$

in complete analogy with the flow of a gas across a normal shock wave.

11   Problems

1. Show that the relation between local and stagnation properties can be expressed in terms of M^* by:

$$\frac{T}{T_0} = \left(1 - \frac{\gamma - 1}{\gamma + 1}M^{*2}\right)$$

$$\frac{\rho}{\rho_0} = \left(1 - \frac{\gamma - 1}{\gamma + 1}M^{*2}\right)^{1/(\gamma-1)}$$

$$\frac{p}{p_0} = \left(1 - \frac{\gamma - 1}{\gamma + 1}M^{*2}\right)^{\gamma/(\gamma-1)}.$$

2. Show that for a perfect gas:

$$M = \frac{u}{a_0}\left[1 - \frac{\gamma - 1}{2}\left(\frac{u}{a_0}\right)^2\right]^{-1/2},$$

and that in the case of low Mach numbers, this equation reduces to:

$$M = \frac{u}{a_0}\left[1 - \frac{\gamma - 1}{2}\left(\frac{u}{a_0}\right)^2\right].$$

3. Show that the ratio between the local temperature of a flow, disturbed by an immersed body, and the free stream temperature is given by:

$$\frac{T}{T_\infty} = 1 - \frac{\gamma - 1}{2}M_\infty^2\left[\left(\frac{u}{U_\infty}\right)^2 - 1\right].$$

4. Show that the maximum velocity that the flow of a perfect gas can attain, leaving a stagnant reservoir, is given by:

$$u_{\max}^2 = 2h_0 = \frac{2}{\gamma - 1}a_0^2.$$

What are the associated temperature and Mach number?

5. Show that for a weak normal shock: $\left(\frac{\Delta p}{p} = \frac{p_2 - p_1}{p_1} \ll 1\right)$:

$$\frac{\Delta \rho}{\rho_1} \approx -\frac{\Delta u}{u_1} \approx \frac{1}{\gamma}\frac{\Delta p}{p_1}$$

$$M_1^2 = 1 + \frac{\gamma + 1}{2\gamma} \frac{\Delta p}{p_1} \qquad \text{(exact)}$$

$$M_2^2 \approx 1 - \frac{\gamma + 1}{2\gamma} \frac{\Delta p}{p_1}$$

$$\frac{\Delta p_0}{p_0} \approx \frac{\gamma + 1}{12\gamma^2} \left(\frac{\Delta p}{p_1}\right)^3.$$

6. Consider the isothermal flow of a perfect gas with $\gamma = 1.4$ in a duct with constant transversal area and friction. Show that the ratio between stagnation pressure at the pipe entrance, where the Mach number is M, and the stagnation pressure p_0^+, at the critical length l^+, is given by:

$$\frac{p_0}{p_0^+} = \frac{1}{\sqrt{\gamma} M} \left\{ \frac{1 + [(\gamma - 1)/2] M^2}{1 + (\gamma - 1)/(2\gamma)} \right\}^{\gamma/(\gamma-1)}.$$

Chapter 3
Oblique Shocks

1 Introduction

This chapter deals with aspects of oblique shock waves, including the approach consisting of describing the shock in terms of the upstream Mach number, the flow deflection, and the shock angle. The concepts of weak and strong shocks are presented. Flows over corners and wedge are discussed. Conditions for the existence of Detached shocks are discussed. As an alternative approach, oblique shocks are discussed in terms of the upstream Mach number and of the downstream velocity components in the directions parallel and perpendicular to the incoming flow. The chapter ends with a discussion of Riemann problems.

Steady two-dimensional oblique shock waves are generated by supersonic flows over corners and wedges [5, 6, 9, 10, 12]. We assume a coordinates frame attached to the solid. Small perturbations introduced in the flow close to the body are transmitted to other regions through weak waves. Since we assume that the inflow is supersonic no wave can propagate upstream the vertex of the wedge. The wave system generated at the solid surface travels with the body. We look for the conditions required for the development of a system of waves and identify the geometry of the body in the neighborhood of the waves system. Since that the influence of the body is limited to the region downstream the shock, methods of analysis of the aerodynamic field be made not applicable to subsonic flows can be employed. Concluding the chapter, we present the method for construction of Riemann problems for Euler's equation, describing the flow in shock tubes.

© The Author(s), under exclusive licence to Springer Nature Switzerland AG 2019
J. Pontes et al., *An Introduction to Compressible Flows with Applications*,
SpringerBriefs in Mathematics, https://doi.org/10.1007/978-3-030-33253-2_3

2   Oblique Shock Waves

We identify initially the conditions for the onset of oblique shocks, when a uniform supersonic flow finds a corner or a wedge. The angle θ of the oblique shock is measured with respect to the direction of the upstream flow. This configuration can be addressed with methods applicable to normal shock waves (Sect. 3), by superimposing to the incident flow a velocity component parallel to the shock wave.

We consider the flow through a shock wave forming an angle β with the velocity w_1 of the uniform upstream flow, as shown in Fig. 3.1. The upstream velocity can be decomposed in a component perpendicular to the shock, u_1, and another one, parallel, v. The incident angle is given by $\beta = \tan^{-1} u_1/v$. The parallel component traverses unaffected the shock, while the perpendicular one undergoes a shock and is abruptly reduced to u_2. Consequently, the outflow velocity emerges from the shock at an angle θ, with the inflow velocity. The flow is always deflected when crossing an oblique shock wave.

Relations between properties at both sides of the shock wave may thus be easily obtained, since super-imposition of a velocity component v, to the component perpendicular to the shock does not change the static pressure and other static parameters at both sides. The Mach number is defined by $M_1 = w_1/a_1$ and the u_1 component, which undergoes the shock, is given by $u_1 = w_1 \sin \beta$. Relations between the properties at both sides are thus obtained from Eqs. 2.32, 2.33, 2.35, and 2.36, upon replacing u_1/a_1 by $M_1 \sin \beta$. We obtain:

$$\frac{\rho_2}{\rho_1} = \frac{u_1}{u_2} = \frac{(\gamma + 1)M_1^2 \sin^2 \beta}{2 + (\gamma - 1)M_1^2 \sin^2 \beta}. \tag{3.1}$$

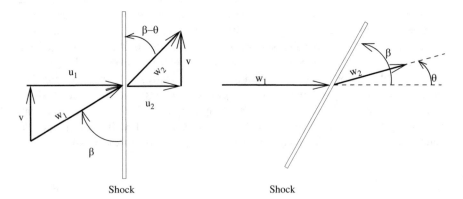

Fig. 3.1 Flow across an oblique shock wave. Left: shock wave with super-imposition of a velocity component v, parallel to the shock. The existence of velocity component parallel to the shock may be attributed to the use of a coordinates frame moving with velocity v, parallel to the shock, and with opposite sense to the one of v, as indicated in the left diagram. Right: Oblique shock

$$\frac{p_2 - p_1}{p_1} = \frac{2\gamma}{\gamma + 1} \left(M_1^2 \sin^2 \beta - 1 \right), \tag{3.2}$$

$$\frac{T_2}{T_1} = 1 + \frac{2(\gamma - 1)}{(\gamma + 1)^2} \frac{\gamma M_1^2 \sin^2 \beta + 1}{M_1^2 \sin^2 \beta} \left(M_1^2 \sin^2 \beta - 1 \right) \tag{3.3}$$

$$M_2^2 \sin^2 (\beta - \theta) = \frac{2 + (\gamma - 1) M_1^2 \sin^2 \beta}{2\gamma M_1^2 \sin^2 \beta - \gamma + 1} \tag{3.4}$$

$$\frac{s_2 - s_1}{R} = \ln \left\{ \left[1 + 2 \frac{2\gamma}{\gamma + 1} \left(M_1^2 \sin^2 \beta - 1 \right) \right]^{1/(\gamma - 1)} \right.$$

$$\left. \times \left[\frac{(\gamma + 1) M_1^2 \sin^2 \beta}{2 + (\gamma - 1) M_1^2 \sin^2 \beta} \right]^{-\gamma/(\gamma - 1)} \right\}. \tag{3.5}$$

The above equations show that the relations between properties up and downstream the shock wave depend only on component of the inflow velocity perpendicular to the shock. This component must be supersonic, i.e., $M_1 \sin \beta \geq 1$, what defines a minimum inclination of the inflow velocity, with respect to the shock. The maximum inclination is $\pi/2$. The inclination angle of an inflow crossing an oblique shock wave must be in the range:

$$\sin^{-1} \frac{1}{M_1} \leq \beta \leq \frac{\pi}{2}. \tag{3.6}$$

The Mach number M_2 downstream the shock is obtained by noting that $M_2 = w_2/a_2$, and that $u_2/a_2 = M_2 \sin (\beta - \theta)$. By replacing the above relations in Eq. 2.31 we have:

$$M_2^2 \sin^2 (\beta - \theta) = \frac{2 + (\gamma - 1) M_1^2 \sin^2 \beta}{2\gamma M_1^2 \sin^2 \beta - \gamma + 1}. \tag{3.7}$$

Each value of the incidence angle β corresponds to a deflection angle θ. β and θ are related by noting that:

$$\tan \beta = \frac{u_1}{v} \quad \text{and} \quad \tan (\beta - \theta) = \frac{u_2}{v}.$$

Eliminating the component v of the velocity from the above two equations and using Eq. 3.1 we find:

$$\frac{\tan (\beta - \theta)}{\tan \beta} = \frac{u_2}{u_1} = \frac{\rho_1}{\rho_2} = \frac{(\gamma - 1) M_1^2 \sin^2 \beta + 2}{(\gamma + 1) M_1^2 \sin^2 \beta}. \tag{3.8}$$

Rearranging terms:

$$\tan\theta = 2\cot\beta\,\frac{M_1^2\sin^2\beta - 1}{M_1^2\left(\gamma + \cos 2\beta\right) + 2}.\tag{3.9}$$

The last expression vanishes for $\beta = \pi/2$ and $\beta = \sin^{-1}1/M_1$, which are the limit values taken by β, as defined by Eq. 3.6. The angle θ is positive in this range and must thus attain a maximum value. For values of $\theta < \theta_{max}$ and of the Mach number M_1 two solutions exist, with different values of β. The largest one defines a *strong shock*, in which the downstream flow is subsonic. The smallest value of β defines an *weak shock*, where the downstream flow is supersonic, except in small range of θ, slightly smaller than θ_{max}.

The relation between β and θ can be written in a useful alternative form, by rearranging Eq. 3.8. We divide both the numerator and the denominator of the expression of the right-hand side of this equation by $M_1^2\sin^2\beta$ to obtain:

$$\frac{1}{M_1^2\sin^2\beta} = \frac{\gamma + 1}{2}\frac{\tan\left(\beta - \theta\right)}{\tan\beta} - \frac{\gamma - 1}{2}$$

or either:

$$M_1^2\sin^2\beta - 1 = \frac{\gamma + 1}{2}M_1^2\frac{\sin\beta\sin\theta}{\cos\left(\beta - \theta\right)}.\tag{3.10}$$

This equation may be rewritten, in an approximate form, for small angles θ, as:

$$M_1^2\sin^2\beta - 1 = \left(\frac{\gamma + 1}{2}M_1^2\tan\beta\right)\theta.\tag{3.11}$$

For high values of M_1, we have $\beta \ll 1$, but $M_1\beta \gg 1$. Equation 3.11 reduces to:

$$\beta \simeq \frac{\gamma + 1}{2}\theta.$$

The angle β of oblique shocks developed when a supersonic flow is deflected by an angle θ and emerges from the shock with Mach number M_2 is graphically represented in Fig. 3.2.

3   The Shock Polar Equation

Section 2 analyzed oblique shocks in terms of the deflection and shock angles, θ and β, respectively, and the upstream Mach number M_1 (see Fig. 3.1). The phenomena can also be studied in terms of the incident critical Mach number M_1^* and the two

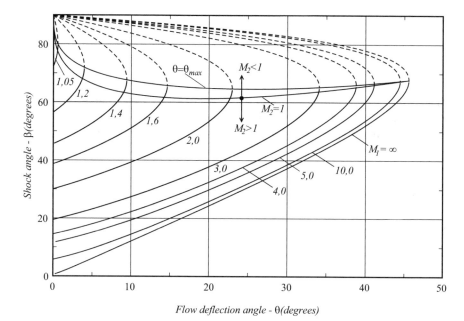

Fig. 3.2 Angle β of the oblique shock wave, relative to the uniform supersonic upstream flow, and developed when the flow with Mach number M_1 is deflected by an angle θ and emerges from the shock with Mach number M_2. The angle β depends on M_1 and θ. The phenomena occurring around an oblique shock wave may be interpreted as those developed when a supersonic flow over a surface defined by two semi planes intercepting at an angle θ is deflected by that angle at the vertex of the corner. An oblique shock develops attached to the corner vertex, if $\theta < \theta_{\max}$. Supersonic flows over corners with $\theta > \theta_{\max}$ develop a curved shock wave, detached from the vertex of the corner, as schematically shown in Fig. 3.4. Dashed lines refer to *strong shocks*, where the deflection angle is smaller than $\theta_{\max}$ and the downstream flow is subsonic. Continuous lines refer to *weak shocks*, where the downstream flow is supersonic except in a small region delimited by the curves of $\theta = \theta_{\max}$ and $M_2 = 1$

components of the flow leaving the shock, w_{2x} and w_{2y}, parallel and perpendicular to the incident flow, respectively. Use of M_1^* instead of M_1 has the advantage that $M_1^* \to 2.45$ as $M_1 \to \infty$.

Since the component of incident flow parallel to the shock is not affected through the discontinuity, we have: $w_1 \cos \beta = w_{2x} \cos \beta + w_{2y} \sin \beta$, or:

$$\tan \beta = \frac{w_1 - w_{2x}}{w_{2y}}.$$

Using Eq. 3.1 we obtain:

$$\frac{w_{2x} \sin \beta - w_{2y} \cos \beta}{w_1 \sin \beta} = \frac{\gamma - 1}{\gamma + 1} + \frac{2a_1^2}{(\gamma + 1) v_1^2 \sin^2 \beta}. \tag{3.12}$$

The angle β can be eliminated from the above equation after some manipulation of terms, leading to the following equation, which relates the components of the downstream velocity, w_{2x} and w_{2y}, respectively, for prescribed upstream velocity w_1, and sound velocity a_1:

$$w_{2y}^2 = (w_1 - v_{2x})^2 \frac{2 \left(w_1 - a_1^2/w_1\right)/(\gamma - 1) - (w_1 - v_{2x})}{w_1 - v_{2x} + 2a_1^2/[(\gamma + 1) v_1]}.$$

From Eq. 2.26 we have:

$$a^{*2} = \frac{2a_1^2 + u_1^2 (\gamma - 1)}{\gamma + 1}.$$

Combination of the two last equations leads to the sought relation between w_{2x} and w_{2y}, in terms of w_1 and a^*:

$$w_{2y}^2 = (w_1 - w_{2x})^2 \frac{w_1 w_{2x} - a^{*2}}{2w_1^2/(\gamma + 1) - w_1 w_{2x} + a^{*2}}. \tag{3.13}$$

This equation can be rewritten in nondimensional form by re-normalizing the velocity components w_1, w_{2x}, and w_{2y} with the critical sound velocity a^*:

$$\bar{w}_1 = \frac{w_1}{a^*} = M_1^*. \qquad \bar{w}_{2x} = \frac{w_{2x}}{a^*} \qquad \bar{w}_{2y} = \frac{w_{2y}}{a^*}.$$

Upon introducing the above definitions in Eq. 3.13 and dropping the bar in the nondimensional variables we obtain the *shock polar equation*:

$$w_{2y}^2 = \frac{\left(M_1^* - w_{2x}\right)^2 \left(w_{2x} M_1^* - 1\right)}{\frac{2}{\gamma + 1} M_1^{*2} - w_{2x} M_1^* + 1}. \tag{3.14}$$

Figure 3.3 shows a plot of Eq. 3.14 with $w_{2y} = f\left(w_{2x}, M_1^*\right)$, for three values of M_1. It is a cubic curve, known as *strophoid*. The strophoid degenerates into a circumference, as $M_1 \to \infty$ ($M_1^* \to 2.45$). A straight line OAB, drawn from the origin and intercepting the strophoid at points A and B, defines the two possible downstream velocities w_2, for prescribed M_1^*. The angle θ, which the line OAB forms with the abscissa axis is the angle the flow is deflected, and the lengths of the segments OA and OB are the two possible values of M_2^*, for prescribed M_1^* and

Fig. 3.3 Shock polar plot for $M_1 = 2$, 5, and ∞, according to Eq. 3.14 (Strophoid curves). Straight lines drawn from the origin intercept the curves at the possible nondimensional pairs (w_{1x}, w_{2x}), for prescribed values of M_1^*. The lengths of segments OA and OB are the absolute values of nondimensional w_2. The curves degenerate into circumferences for $M_1 \to \infty$ ($M^* \to 2.45$)

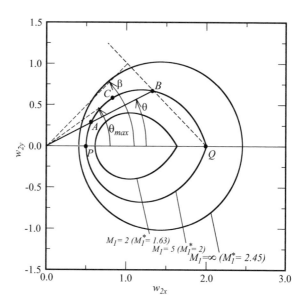

deflection angle. The plot shows that a limit deflection angle θ_{max} exists, beyond which no attached shocks are possible. The same result was obtained in Sect. 2. The polar shock curves cross the abscissaæ axe at points P and Q, where the nondimensional $w_{2x} = M_{2*}$ and $w_{2y} = 0$. At both points $w_{2x} = M_2^* = 1/M_1^*$ but at the point Q, $M_1^* = M_2^* = 1$. In both cases, the deflection angle is zero, namely the shock is normal to the incident flow. In the first case (point P) the pressure recovery along the shock is maximum, according to Eq. 3.2. In the second one (point Q), no shock occurs, and the line defined by the shock is just a characteristic.

From Fig. 3.3 we see that $w_2 \cos(\beta - \theta) = w_{2\,max} \cos \beta$ and to the fact that $w_{2\,max}$ is parallel to w_1 it is clear that the angle β of the shock wave with respect to the incident flow is the one defined by the perpendicular to the straight lines QB or QA drawn from the origin, and the direction of w_1.

The strophoids actually continue in two branches for $w_{2x} > w_{1x}$, with $w_{2y} \to \infty$, up to a common asymptote given by:

$$w_{2x} = \frac{2M_1^*}{\gamma + 1} + \frac{1}{M_1^*}.$$

The two branches above mentioned are associated to downstream Mach numbers larger than the upstream one, which are not physical.

As θ decreases the point A approaches P, corresponding to a normal shock, with $\beta = \pi/2$, with $w_2 = a^{*2}/w_1$, and the point B approaches Q, with $w_{2x} \to w_1$. The shock intensity, characterized by the velocity discontinuity, tends to zero.

The two possible shock waves associated with the same deflection angle θ belong to either the class of *strong shocks*, with large wave angle β and downstream subsonic velocity, or to the class of *weak shocks*, with smaller shock angle β and downstream almost always supersonic velocity.

4 Supersonic Flows Over Corners and Wedges

If we consider inviscid flows, stream lines can always be replaced by solid walls. In consequence, supersonic flows crossing an oblique shock, as described in Sect. 2, are a solution of flows over wedges and corners (see Fig. 3.4).

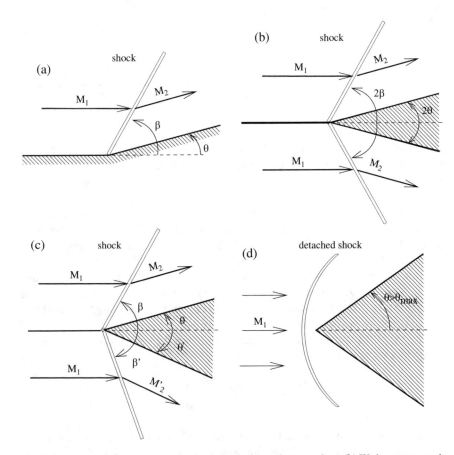

Fig. 3.4 Supersonic flow over corners, as shown in (**a**), and over wedges. (**b**) Wedge at zero angle of attack and $\theta < \theta_{max}$. An oblique shock develops, attached to the wedge vertex. (**c**) Wedge with nonzero angle of attack, and angles θ and $\theta' < \theta_{max}$. Again, the flow develops a shock attached to the vertex. (**d**) Wedge with $\theta > \theta_{max}$. The flow develops a curved shock wave, detached from the wedge vertex

The deflection angle θ of the corner defines two possible shock angles β and the corresponding downstream Mach number M_2. We only consider *weak shocks*, where $M_2 > 1$, a restriction that implies in $\theta < \theta_{max}$. The flow over a wedge with faces at angle 2θ, and zero angle of attack is obtained by symmetry. The aerodynamic fields above and below the symmetry line of the wedge are independent. Configurations where the symmetry line is at a nonzero angle of attack with the uniform upstream flow are also possible, as shown in Fig. 3.4.

5 Riemann Problems for the Euler Equation

This section presents the construction of the so-called Riemann's problem for Euler's equation. Riemann's problem consists of building an initial value problem for conservation laws, when the initial condition consists of two constant states U_L and U_R, separated by a steep discontinuity at $x = 0$.

Riemann's problem for Euler's equation in one dimension addresses the phenomena occurring in a *shock tube*, when a membrane separating a tube in two regions is removed. Each side of the tube contains initially a gas where pressure and specific mass are initially uniform and constant, but with different values at the two sides. Removing the membrane gives raise to waves traveling in both sides, described by the solution of Riemann's problem. Riemann's problem is used in the construction of more general solutions for conservation laws and in the construction of numerical methods for the simulation of compressible flows.

Simple waves provide a solution of Riemann's problem in regions of expansion waves, or *rarefaction waves*. In regions where compression occurs, solutions of Riemann's problem are in the form of *shock waves*.

We consider Riemann's problem for Euler's equation in one dimension. The problem is stated as:

$$\frac{\partial \mathbf{U}}{\partial t} + \frac{\partial \mathbf{F}}{\partial x} = 0; \qquad \mathbf{U}(x,0) = \mathbf{U}_0(x), \tag{3.15}$$

where the vector of conservative flow variables $\mathbf{U}$ and $\mathbf{F}$ is given by:

$$\mathbf{U} = \rho \begin{Bmatrix} 1 \\ u \\ E \end{Bmatrix}, \quad \mathbf{F} = \rho u \begin{Bmatrix} 1 \\ u \\ E \end{Bmatrix} + p \begin{Bmatrix} 0 \\ 1 \\ u \end{Bmatrix}, \tag{3.16}$$

where ρ is the density, u is the fluid velocity, and E is the total energy per unit volume

$$E = e + \frac{u^2}{2}, \tag{3.17}$$

with Riemann's initial conditions given by:

$$\mathbf{U}_0(x) = \begin{cases} U_L, \text{ for } x < 0, \\ U_R, \text{ for } x > 0. \end{cases} \tag{3.18}$$

It was shown that, in the case of the wave equation (Eq. 1.9, Chap. 1) two families of characteristics exist along the straight lines defined by $\lambda = dx/dt = \pm c$. In the case of the Euler's equation for compressible flows, the aerodynamic field can be classified into three characteristic regions [8]:

$$\lambda_1 = u - c, \quad \lambda_2 = u, \quad \lambda_3 = u + c.$$

In order to prove this statement we rewrite Euler's equation as:

$$\frac{\partial \mathbf{U}}{\partial t} + \mathbf{A}\frac{\partial \mathbf{U}}{\partial x} = 0 \quad \mathbf{U}(x,0) = \mathbf{U}_0(x), \tag{3.19}$$

where $A_{ij} = \partial F_i / \partial U_j$ (see Problems 5 and 6). This system can be diagonalized by left multiplying the system by $\mathbf{L}$, the matrix of left eigenvectors of $\mathbf{A}$, and therefore Eq. 3.15 can be written as

$$\mathbf{L}\frac{\partial \mathbf{U}}{\partial t} + \mathbf{L}A\frac{\partial \mathbf{U}}{\partial x} = 0 \tag{3.20}$$

or

$$\mathbf{L}\frac{\partial \mathbf{U}}{\partial t} + \Lambda\mathbf{L}\frac{\partial \mathbf{U}}{\partial x} = 0, \tag{3.21}$$

where

$$\Lambda = \begin{bmatrix} \lambda_1 & 0 & 0 \\ 0 & \lambda_2 & 0 \\ 0 & 0 & \lambda_3 \end{bmatrix}. \tag{3.22}$$

It can be shown that $\mathbf{V} = \mathbf{L}\mathbf{U}$ satisfies the equation:

$$\frac{\partial \mathbf{V}}{\partial t} + \Lambda\frac{\partial \mathbf{V}}{\partial x} = 0, \tag{3.23}$$

namely, $\mathbf{V}$ is constant along the characteristic curves and is denoted as the Riemann's invariant.

The fields λ_1 and λ_3 are strictly nonlinear and develop shock/rarefaction nonlinear waves, whereas λ_2 is linearly degenerated and develops only one contact wave.

5.1 Shocks

The following relations hold for a shock wave:

$$p - p_0 = \pm \rho_0 c_0 \sqrt{1 + \frac{\gamma + 1}{2\gamma}\left(\frac{p}{p_0} - 1\right)}(u - u_0) \qquad (3.24)$$

$$\frac{\rho}{\rho_0} = \frac{1 + \dfrac{\gamma + 1}{\gamma - 1}\dfrac{p}{p_0}}{\dfrac{\gamma + 1}{\gamma - 1} + \dfrac{p}{p_0}}. \qquad (3.25)$$

The above relations can be derived from the Rankine-Hugoniot relations. The shock velocity is given by:

$$V_s = u_0 \pm c_0 \sqrt{1 + \frac{\gamma + 1}{2\gamma}\left(\frac{p}{p_0} - 1\right)}. \qquad (3.26)$$

5.2 Rarefactions

In the case of rarefaction waves, the following relations are valid:

$$u \pm \frac{2}{\gamma - 1}c = u_0 \pm \frac{2}{\gamma - 1}c_0, \qquad (3.27)$$

$$p - p_0 = \pm \rho_0 c_0 \frac{\gamma - 1}{2\gamma}\frac{1 - (p/p_0)}{1 - (p/p_0)^{\frac{\gamma - 1}{2\gamma}}}(u - u_0). \qquad (3.28)$$

The above relations are obtained from the isentropic equations and the Riemann invariant $u + \frac{2c}{\gamma - 1} = $ constant. Equation 3.28 defines a curve in the space (p, u) denoted as *Poisson's curve*. Through the rarefaction wave, u and c vary linearly across the rarefaction wave, and p and ρ given by:

$$p = p_o \frac{c}{c_0}^{\frac{2\gamma}{\gamma - 1}}$$

$$\rho = \rho_o \frac{p}{p_0}^{\frac{1}{\gamma}}.$$

Construction of a Riemann's problem can be made by following a sequence of steps. For instance, construction of a problem admitting a solution in the form of a shock

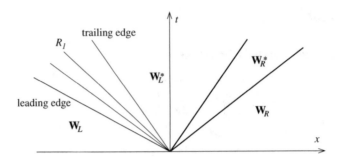

Fig. 3.5 Construction of a Riemann problem: a shock tube. Shock λ_3 S_3; Contact C_2, and Rarefaction λ_1 (R_1)

and a contact wave traveling to the right and a rarefaction wave traveling to the left (see Fig. 3.5) can be done as follows:

1. Prescribe the right-hand side $\mathbf{W}_R = [\rho_R, u_R, p_R]^t$.
2. Prescribe p^* ($> p_R$), and determine u^* with Eq. 3.24, namely:

$$u^* = u_R + \frac{p^* - p_R}{\rho_R c_R \sqrt{1 + \frac{\gamma+1}{2\gamma}\left(\frac{p^*}{p_R} - 1\right)}}. \tag{3.29}$$

Determine ρ_R^* by Eq. 3.25, i.e.,

$$\rho_R^* = \rho_R \frac{1 + \dfrac{\gamma + 1}{\gamma - 1}\dfrac{p^*}{p_R}}{\dfrac{\gamma + 1}{\gamma - 1} + \dfrac{p^*}{p_R}}. \tag{3.30}$$

3. Prescribe ρ_L^*. Velocity and pressure are not affected across the contact: $u_L^* = u_R^* = u^*$ and $p_L^* = p_R^* = p^*$.
4. Prescribe p_L($> p^*$), and determine ρ_L with

$$\rho_L = \rho_L^* \frac{p_L^{\frac{1}{\gamma}}}{p^*}$$

and the velocity with Eq. 3.28.

$$u_L = u^* + \frac{2\gamma}{(\gamma - 1)\rho_L c_L} \frac{1 - (p^*/p_L)^{\frac{\gamma-1}{2\gamma}}}{1 - (p^*/p_L)}(p^* - p_L). \tag{3.31}$$

The above steps lead to a well-defined Riemann's problem. The shock velocity associated with this problem is given by:

$$V_{S3} = u_R + c_R \sqrt{1 + \frac{\gamma + 1}{2\gamma} \left(\frac{p^*}{p_R} - 1 \right)}, \tag{3.32}$$

the contact velocity is given by

$$V_{C2} = u^*, \tag{3.33}$$

and the leading and trailing velocity of the rarefaction by:

$$V_{R1}^{tail} = u_L^* - c_L^* \tag{3.34}$$

$$V_{R1}^{front} = u_L - c_L. \tag{3.35}$$

The above velocities can be of use when choosing a convenient shock velocity. To find a solution inside a rarefaction wave, we initially evaluate the velocity, which varies linearly, and then, the remaining variables:

$$u(\xi) = (1 - \xi)u_L + \xi u^* \tag{3.36}$$

$$c(\xi) = (1 - \xi)c_L + \xi c^* \tag{3.37}$$

$$p(\xi) = p_L \left(\frac{c(\xi)}{c_L} \right)^{\frac{2\gamma}{\gamma - 1}} \tag{3.38}$$

$$\rho(\xi) = \rho_L \left(\frac{p(\xi)}{p_L} \right)^{\frac{1}{\gamma}}, \tag{3.39}$$

where $\xi = (x - x_{Front})/(x_{Tail} - x_{Front})$. As can be observed, it is possible to build several Riemann problems. It is possible, for instance, after calculating $\mathbf{W}_R^*$, to specify $\mathbf{W}_L = \mathbf{W}_R^*$, which defines a Riemann problem with a unique shock, or if $\rho_L^* = \rho_R^*$, there are no contact waves. Exact solutions for Riemann problems can be employed as benchmarks for validation of computational methods. Figure 3.6 illustrates the exact solution of the associated Riemann problem for a shock tube, build with $\mathbf{W}_R = [\rho_R, u_R, p_R]^t = [0.125, 0, 0.1]^t$, and $\mathbf{W}_L = [\rho_L, u_L, p_L]^t = [1, 0, 1]^t$. We stress that the solution is *similar*, so $\mathbf{W}(x, t) = \mathbf{W}(\eta)$, where $\eta = x/t$.

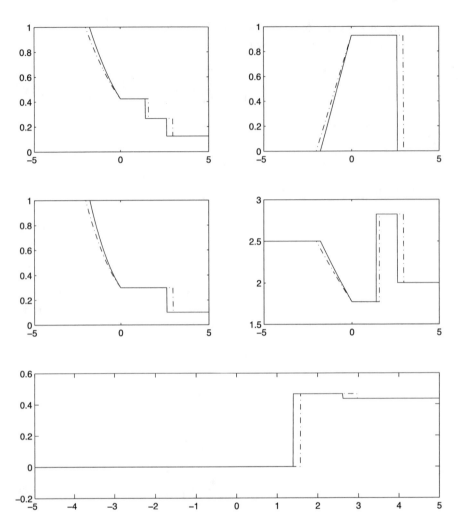

Fig. 3.6 Exact solution of a Riemann's problem associated with a shock tube, built with $\mathbf{W}_R = [\rho_R, u_R, p_R]^t = [0.125, 0, 0.1]^t$, and $\mathbf{W}_L = [\rho_L, u_L, p_L]^t = [1, 0, 1]^t$. Plots refer to ρ, u, p, $e - u^2/2$, e $1/(\gamma - 1)\log(p/\rho^\gamma)$, at $t = 1.5$ (solid line) and at $t = 1.7$ (dashed line)

6   Problems

1. Show that the deviation angle θ of the velocity in an oblique shock wave occurring in the flow of a perfect gas is related to the angle β between the shock wave and the direction of the incident velocity w_1 and to the incident Mach number M_1 through the formula:

$$\cot \theta = \tan \beta \left[\frac{(\gamma + 1) M_1^2}{2 \left(M_1^2 \sin^2 \beta - 1 \right)} - 1 \right].$$

2. Show that the flow Mach number M_2 leaving an oblique shock wave relates to the incident Mach number M_1 and to the angle of the downstream flow with the direction of the incident flow through the formula:

$$M_2^2 = \frac{2 + (\gamma - 1) M_1^2}{2\gamma M_1^2 \sin^2 \beta - (\gamma - 1)} + \frac{2M_1^2 \cos^2 \beta}{2 + (\gamma - 1) M_1^2 \sin^2 \beta}.$$

3. Show that the pressure rate across an oblique shock is given by the following relation, for prescribed M_1 and β:

$$\frac{p_2}{p_1} = \frac{2\gamma M_1^2 \sin^2 \beta - (\gamma - 1)}{\gamma + 1}.$$

4. Show that the components w_{2x} and w_{2y} of the velocity leaving an oblique shock wave are related by the following equation when the upstream Mach number $M_1 \to \infty$:

$$w_{2y}^2 = \frac{[(\gamma + 1)/(\gamma - 1) - w_{2x}][(\gamma + 1)/(\gamma - 1) w_{2x} - 1]}{(\gamma + 1)/(\gamma - 1)},$$

which is the equation of a circumference, in accordance with Fig. 3.3.
5. Show that $A_{ij} = \partial F_i / \partial U_j$ in Eq. 3.15 is given by

$$A = \begin{bmatrix} 0 & 1 & 0 \\ (\gamma - 3)\frac{u^2}{2} & (3 - \gamma)u & \gamma - 1 \\ \left(\frac{\gamma-1}{2}\right)u & H + (1 - \gamma)u^2 & \gamma u \end{bmatrix}, \tag{3.40}$$

where $H = E + \frac{p}{\rho}$.
6. Show that Eq. 3.19 can be diagonalized by left multiplying the system by L, the matrix of left eigenvectors of A, such that:

$$L dU = \begin{bmatrix} \frac{dp - \rho c \, du}{2c} \\ -\frac{dp - \rho c^2 d\rho}{2c} \\ \frac{dp + \rho c \, du}{c^2} \end{bmatrix}. \tag{3.41}$$

7. **Solution of one-dimensional hyperbolic conservation equations in the time domain through the Finite Differences Method**
This problem discusses the solution of one-dimensional hyperbolic conservation equations in the time domain through the Finite Differences Method and illustrate the method with a computational example using the MacCormack's method with addition of a Lapidus' artificial viscosity term.

The MacCormack's Method

MacCormack's method features a two-step second order expressed by:

$$\mathbf{U}_i^{\overline{n+1}} = \mathbf{U}_i^n - \frac{\Delta t}{\Delta x}\left(\mathbf{F}_{i+1}^n - \mathbf{F}_i^n\right)$$

$$\tilde{\mathbf{U}}_i^{n+1} = \frac{1}{2}\left(\mathbf{U}_i^n + \mathbf{U}_i^{\overline{n+1}}\right) - \frac{\Delta t}{2\Delta x}\left(\mathbf{F}_i^{\overline{n+1}} - \mathbf{F}_{i-1}^{\overline{n+1}}\right).$$

The above two steps define MacCormack's method. Usually, a term is added to provide an artificial viscosity, with an extra fractional step, as proposed by Lapidus to stabilize the scheme:

$$\mathbf{U}_i^{n+1} = \tilde{\mathbf{U}}_i^{n+1} + \frac{\nu\Delta t}{2\Delta x}\Delta'\left(\|\Delta'\tilde{\mathbf{U}}_i^{n+1}\| \cdot \Delta'\tilde{\mathbf{U}}_i^{n+1}\right),$$

where $\Delta'\tilde{\mathbf{U}}_i^n = \tilde{\mathbf{U}}_i^n - \tilde{\mathbf{U}}_{i-1}^n$. An implementation of the method using the artificial viscosity of Lapidus is given below:

```
clear;clc;
% grid generation
nx=1000;
Lx=10;dx=Lx/(nx-1);
X=0:dx:Lx;
% simulation parameters
nu=4.0;
gamma=1.4;
% simulation time
tend=0.5;
dt = 0.0005;
% initial conditions
% -- rho (density)
rho=ones(nx,1);
rho(fix(nx/2):end,1)=0.125;
% -- rho * u
rhou=zeros(nx,1);
% -- pressure
p = ones(nx,1);
p(fix(nx/2):end,1) = 0.1;
% -- total energy rhoe
rhoe = p./(gamma-1) + 0.5*(rhou.*rhou./rho);

% operators assembling
Dxp=zeros(nx,nx); % forward difference
Dxr=zeros(nx,nx); % backward difference
for i=2:nx-1
  % forward
  Dxp(i,i+1)=1/dx;
  Dxp(i,i)=-1/dx;
  % backward
  Dxr(i,i)=1/dx;
  Dxr(i,i-1)=-1/dx;
end
```

```
% time step loop
t=0; % initial time t
while t<tend
 % first step
 rhostar = rho - dt*(Dxp*rhou);
 rhoustar = rhou - dt*(Dxp*(rhou.*rhou./rho + p));
 rhoestar = rhoe - dt*(Dxp*( (rhou./rho).*(rhoe+p) ));
 pstar = (gamma-1.0)*(rhoestar - 0.5*(rhoustar.*rhoustar./rhostar) );

 % second step
 rho =  0.5*(rho+rhostar) -(0.5*dt)*( Dxr*rhoustar );
 rhou = 0.5*(rhou+rhoustar)...
         -(0.5*dt)*(Dxr*(rhoustar.*rhoustar./rhostar + pstar) );
 rhoe = 0.5*(rhoe+rhoestar)...
       -(0.5*dt)*( Dxr*( (rhoustar./rhostar).*(rhoestar+pstar) ));

 % third step (Lapidus) - optional
 rho = rho +(dt*dx*dx*nu)*( Dxr*(abs(Dxp*rho).*(Dxp*rho)) );
 rhou = rhou +(dt*dx*dx*nu)*( Dxr*(abs(Dxp*rhou).*(Dxp*rhou)) );
 rhoe = rhoe +(dt*dx*dx*nu)*( Dxr*(abs(Dxp*rhoe).*(Dxp*rhoe)) );
 p = (gamma-1.0)*( rhoe - 0.5*(rhou.*rhou./rho) );

 % computing e (internal energy)
 ei = rhoe./rho - 0.5*(rhou.*rhou./rho./rho);
  % computing u (velocity)
 u = rhou./rho;
  % computing s (entropy)
 s = 1/(gamma-1)*log(p./(rho.^gamma));
 % incrementing time t
 t=t+dt
end

% calling plot function defined below
plotfun(X,u,p,ei,gamma,rho,s,nx)

% %% ----------- Plotting figures using subplot --------- %%%
function a=plotfun(X,u,p,ei,gamma,rho,s,nx)
 % plotting density \rho
 figure(1)
 subplot(3,2,1)
 plot(X,rho,'k-','LineWidth',2)
 xlabel('x','FontSize', 16)
 xtickformat('%.1f')
 ylabel('$\rho$','FontSize',16)
 ytickformat('%.1f')
 set(gca,'FontSize',18)

 % plotting velocity u
 subplot(3,2,2)
 plot(X,u,'k-','LineWidth',2)
 xlabel('x','FontSize', 16)
 xtickformat('%.1f')
 ylabel('u','FontSize',16)
```

```
ytickformat('%.1f')
set(gca,'FontSize',18)

% plotting pressure p
subplot(3,2,3)
plot(X,p,'k-','LineWidth',2)
xlabel('x','FontSize', 16)
xtickformat('%.1f')
ylabel('p','FontSize',16)
ytickformat('%.1f')
set(gca,'FontSize',18)

% plotting internal energy ei
subplot(3,2,4)
plot(X,ei,'k-','LineWidth',2)
xlabel('x','FontSize', 16)
xtickformat('%.1f')
ylabel('ei','FontSize',16)
ylim([1.5 3.1])
%set(gca,'YLim',[1.5 3.0])
set(gca,'YTick',[1.5:0.5:3.0])
ytickformat('%.1f')
set(gca,'FontSize',18)

% plotting entropy s
subplot(3,2,[5 6])
plot(X,s,'k-','LineWidth',2)
xlabel('x','FontSize', 16)
xtickformat('%.1f')
ylabel('s','FontSize',16)
ylim([-0.2 2.2])
ytickformat('%.1f')
set(gca,'YTick',[0.0:0.5:2.0])
set(gca,'FontSize',18)

drawnow;
end
```

Validation of the above methodology can be made by evaluating the solution of the pressure and the velocity fields in a shock tube, using the MacCormack's method in a uniform grid with 50 points and a second one with 1000 points, and comparing the results with the analytical solution to confirm the exactness of the method. A result showing the solution at $t = 0.15$ is given in Fig. 3.7.

The Flux Jacobian Method

As can be seen in the previous item, the MacCormack's method introduces spurious oscillating in the solution. Such a spurious oscillating can be dumped out using an additional term to include artificial viscosity as proposed by Lapidus. In this case, the numerical solution does not preserve correctly the stiff steps of the equation solution, smoothing out zones artificially. A better solution may be found by using the linearization of the nonlinear derivative found in the

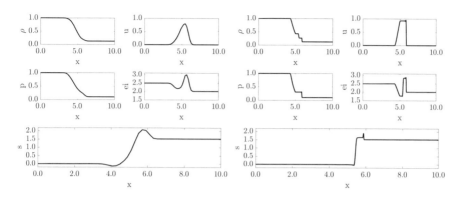

Fig. 3.7 Solution of the pressure and the velocity fields in linear shock tube using the MacCormack's method in a uniform grid with 50 nodes (left) and in grid with 1000 points (right). The figure presents a plot of density ρ, velocity u, pressure p, internal energy $ei = e - u^2/2$, and entropy $s = 1/(\gamma - 1)\log(p/\rho^{\gamma})$ for each case

conservative form of the Euler equations through a flux Jacobian. The set of equations is written in the conservative form as below:

$$\partial \mathbf{U} \partial t + \partial \mathbf{F} \partial x = 0. \tag{3.42}$$

The vectors of conservative variables $\mathbf{U}$ and of fluxes $\mathbf{F}$ are given by:

$$\mathbf{U} = \left\{ \begin{array}{c} p \\ \rho u \\ \rho e \end{array} \right\}, \qquad \mathbf{F} = \left\{ \begin{array}{c} \rho u \\ \rho u^2 + p \\ \rho u H \end{array} \right\}. \tag{3.43}$$

The flux Jacobian is given by:

$$\mathbf{A} = \frac{\partial \mathbf{F}}{\partial \mathbf{U}} = \left\{ \begin{array}{ccc} 0 & 1 & 0 \\ (\gamma - 3)\dfrac{u^2}{2} & (3 - \gamma)u & \gamma - 1 \\ \left(\dfrac{\gamma - 1}{2}u^2 - H\right)u & H + (1 + \gamma)u^2 & \gamma u \end{array} \right\}. \tag{3.44}$$

The eigenstructure is defined as:

$$\mathbf{A} = \mathbf{R}\Lambda\mathbf{L}, \tag{3.45}$$

where:

$$\Lambda = = \left\{ \begin{array}{ccc} u - c & 0 & 0 \\ 0 & u & 0 \\ 0 & 0 & u + c \end{array} \right\}, \tag{3.46}$$

$$\mathbf{R} == \left\{ \begin{matrix} 1 & 1 & 1 \\ u - c & u & u + c \\ H - uc & u^2/2 & H + uc \end{matrix} \right\}, \tag{3.47}$$

$$\mathbf{L} == \left\{ \begin{matrix} \frac{1}{2}\left(\frac{\gamma-1}{2c^2}u^2 + \frac{u}{c}\right) & -\frac{1}{2}\left(\frac{\gamma-1}{c^2}u + \frac{1}{c}\right) & \frac{\gamma-1}{2c^2} \\ 1 - \frac{\gamma-1}{2c^2}u^2 & \frac{\gamma-1}{c^2}u - \frac{1}{c} & -\frac{\gamma-1}{c^2} \\ \frac{1}{2}\left(\frac{\gamma-1}{2c^2}u^2 - \frac{u}{c}\right) & -\frac{1}{2}\left(\frac{\gamma-1}{c^2}u - \frac{1}{c}\right) & \frac{\gamma-1}{2c^2} \end{matrix} \right\}, \tag{3.48}$$

$$\mathbf{R} == \left\{ \begin{matrix} \frac{dp-\rho cdu}{2c^2} \\ -\frac{dp-c^2d\rho}{c^2} \\ \frac{dp+\rho cdu}{2c^2} \end{matrix} \right\}. \tag{3.49}$$

An implementation of the flux Jacobian method is detailed below:

```
% grid generation
nx=50;
Lx=10;dx=Lx/(nx-1);
X=0:dx:Lx;
% simulation parameters
nu=4.0;
gamma=1.4;
% simulation time
tend=0.5;
dt = 0.0005;
% initial conditions
% -- rho
rho=ones(nx,1);
rho(fix(nx/2):end,1)=0.125;
% -- rho * u
rhou=zeros(nx,1);
% -- u (velocity)
u = rhou./rho;
% -- pressure
p = ones(nx,1);
p(fix(nx/2):end,1) = 0.1;
% -- speed of sound
c = sqrt(gamma*p./rho);
% -- total energy rho*e
rhoe = p./(gamma-1) + 0.5*(rhou.*rhou);
% -- specific total enthalpy
H = rhoe./rho + p./rho;
% internal energy ei (e - 1/2 u^2)
ei = rhoe./rho - 0.5*(rhou.*rhou./rho./rho);

% operators assembling
Dxp=zeros(nx,nx); % forward difference
Dxr=zeros(nx,nx); % backward difference
```

```
for i=2:nx-1
 % forward
 Dxp(i,i+1)=1/dx;
 Dxp(i,i)=-1/dx;
 % backward
 Dxr(i,i)=1/dx;
 Dxr(i,i-1)=-1/dx;
end

lambda = zeros(nx,3);
t=0; % initial time t
while t<tend
 c = sqrt(gamma*p./rho);
 lambda = [ u-c u u+c ];
 sn=lambda<0;
 B1 = (lambda(:,1)./(2*c.^2) ).*( sn(:,1).*(Dxp*p-rho.*c.*Dxp*u) + ...
                                 (1-sn(:,1)).*(Dxr*p-rho.*c.*Dxr*u));
 B2 = (lambda(:,2)./(2*c.^2) ).*( sn(:,2).*(-Dxp*p+c.^2.*Dxp*rho) + ...
                                 (1-sn(:,2)).*(-Dxr*p+c.^2.*Dxr*rho));
 B3 = (lambda(:,3)./(c.^2)   ).*( sn(:,3).*(Dxp*p+rho.*c.*Dxp*u) + ...
                                 (1-sn(:,3)).*(Dxr*p+rho.*c.*Dxr*u));
 % mounting Jacobian
 R = zeros(3,3,nx);
 R(1,1,:) = ones(nx,1);
 R(1,2,:) = ones(nx,1);
 R(1,3,:) = ones(nx,1);
 R(2,1,:) = u-c;
 R(2,2,:) = u;
 R(2,3,:) = u+c;
 R(3,1,:) = H - u.*c;
 R(3,2,:) = u.^2/2;
 R(3,3,:) = H + u.*c;

 % updating U=[rho, rhou, rhoe]
 for i=1:nx
    U=[rho(i); rhou(i); rhoe(i)];
    Unew = U - dt*R(:,:,i)*[B1(i); B2(i); B3(i)];
    rho(i)=Unew(1);
    rhou(i)=Unew(2);
    rhoe(i)=Unew(3);
 end
 % computing pressure
 p = (gamma-1.0)*(rhoe - 0.5*(rhou.*rhou./rho) );
 % computing h (entalpy)
 H = rhoe./rho + p./rho;
 % computing ei (internal energy)
 ei = rhoe./rho - 0.5*(rhou./rho).*(rhou./rho);
 % computing u (velocity)
 u = rhou./rho;
 % computing entropy s
 s = 1/(gamma-1)*log(p./(rho.^gamma));
 % incrementing time t
 t=t+dt
end
```

```
% calling plot function defined below
plotfun(X,u,p,ei,gamma,rho,s,nx)

% %% ----------- Plotting figures using subplot --------- %%%
function a=plotfun(X,u,p,ei,gamma,rho,s,nx)
 % plotting density \rho
 figure(1)
 subplot(3,2,1)
 plot(X,rho,'k-','LineWidth',2)
 xlabel('x','FontSize', 16)
 xtickformat('%.1f')
 ylabel('$\rho$','FontSize',16)
 ytickformat('%.1f')
 set(gca,'FontSize',18)

 % plotting velocity u
 subplot(3,2,2)
 plot(X,u,'k-','LineWidth',2)
 xlabel('x','FontSize', 16)
 xtickformat('%.1f')
 ylabel('u','FontSize',16)
 ytickformat('%.1f')
 set(gca,'FontSize',18)

 % plotting pressure p
 subplot(3,2,3)
 plot(X,p,'k-','LineWidth',2)
 xlabel('x','FontSize', 16)
 xtickformat('%.1f')
 ylabel('p','FontSize',16)
 ytickformat('%.1f')
 set(gca,'FontSize',18)

 % plotting internal energy ei
 subplot(3,2,4)
 plot(X,ei,'k-','LineWidth',2)
 xlabel('x','FontSize', 16)
 xtickformat('%.1f')
 ylabel('ei','FontSize',16)
 ylim([1.5 3.0])
 %set(gca,'YLim',[1.5 3.0])
 set(gca,'YTick',[1.5:0.5:3.0])
 ytickformat('%.1f')
 set(gca,'FontSize',18)

 % plotting entropy s
 subplot(3,2,[5 6])
 plot(X,s,'k-','LineWidth',2)
 xlabel('x','FontSize', 16)
 xtickformat('%.1f')
 ylabel('s','FontSize',16)
 ytickformat('%.1f')
 set(gca,'YTick',[0.0:0.5:2.0])
```

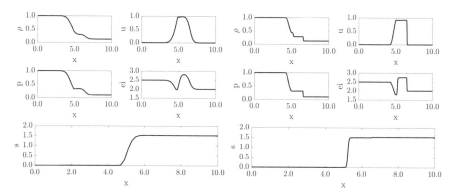

Fig. 3.8 Solution of the pressure and the velocity fields in linear shock tube using the Flux Jacobian method in a uniform grid with 50 nodes (left) and in grid with 1000 points (right). The figure presents a plot of density ρ, velocity u, pressure p, internal energy $ei = e - u^2/2$, and entropy $s = 1/(\gamma - 1)\log(p/\rho^\gamma)$ for each case

```
set(gca,'FontSize',18)

drawnow;
end
```

Validation of the above methodology can be made by evaluating the solution of the pressure and the velocity fields in a shock tube, using the Flux Jacobian method in a uniform grid with 50 points and a second one with 1000 points, and comparing the results with the analytical solution to confirm the exactness of the method. A result, showing the solution at $t = 0.15$ is given in Fig. 3.8.

Chapter 4
Uniform Flows with Small Perturbations

1 Introduction

This chapter addresses the principles of small perturbation theory, where a high speed incoming flow is perturbed by slender bodies aligned or quasi aligned to the flow. Steady state is assumed. The equation of the potential ϕ associated with irrotational compressed flow is simplified for subsonic and supersonic regimes, where the problem is linear. The nonlinear equation governing the behavior of steady three-dimensional transonic flow is presented. A derivation is given for the pressure coefficient. The chapter discusses the problem of two-dimensional steady flow over a periodic-shaped wall both in the linear subsonic and supersonic regimes.

2 The Equations of the Steady Perturbed Flows

We consider the problem where a uniform steady flow moving with velocity U is slightly perturbed by a slender body aligned or almost aligned to the flow. The presence of the body induces the existence of small perturbations u, v, and w, along the directions x, y, and z, respectively. The flow is governed by the potential equation of compressible flows, Eq. 1.5, simplified to account for the hypothesis of small perturbations. Nevertheless, we retain certain nonlinear terms in the potential equation. We here assume that:

$$\frac{u}{U}, \frac{v}{U}, \frac{w}{U} \ll 1, \tag{4.1}$$

and simplify the governing equations, by dropping small nonlinear terms, containing products of the perturbation components.

© The Author(s), under exclusive licence to Springer Nature Switzerland AG 2019
J. Pontes et al., *An Introduction to Compressible Flows with Applications*,
SpringerBriefs in Mathematics, https://doi.org/10.1007/978-3-030-33253-2_4

Steady compressible potential flows are governed by the equation (see Eq. 1.5):

$$\frac{\partial^2 \phi}{\partial x_i^2} = \frac{1}{a^2} \frac{\partial \phi}{\partial x_i} \frac{\partial \phi}{\partial x_j} \frac{\partial^2 \phi}{\partial x_i \partial x_j}. \tag{4.2}$$

Writing down the above equation in all terms, we have:

$$a^2 \left(\frac{\partial u}{\partial x_1} + \frac{\partial v}{\partial x_2} + \frac{\partial w}{\partial x_3} \right) = (U+u)^2 \frac{\partial u}{\partial x_1} + v^2 \frac{\partial v}{\partial x_2} + w^2 \frac{\partial w}{\partial x_3}$$

$$+ v \, (U+u)^2 \left(\frac{\partial u}{\partial x_2} + \frac{\partial v}{\partial x_1} \right) + vw \left(\frac{\partial v}{\partial x_3} + \frac{\partial w}{\partial x_2} \right)$$

$$+ w \, (U+u)^2 \left(\frac{\partial w}{\partial x_1} + \frac{\partial u}{\partial x_3} \right). \tag{4.3}$$

From Eq. 2.7 we have:

$$\frac{a_\infty^2}{\gamma - 1} + \frac{U^2}{2} = \frac{a^2}{\gamma - 1} + \frac{(U+u)^2 + v^2 + w^2}{2}$$

or:

$$a^2 = a_\infty^2 - \frac{\gamma - 1}{2} \left(2uU + u^2 + v^2 + w^2 \right). \tag{4.4}$$

Substituting the last result in Eq. 4.3 and rearranging terms, we obtain the full nonlinear equation of the potential flow, in terms of the velocity components, and applicable to an incident uniform flow, perturbed by the presence of a slender body aligned or quasi aligned to the flow:

$$\left(1 - M_\infty^2 \right) \frac{\partial u}{\partial x_1} + \frac{\partial v}{\partial x_2} + \frac{\partial w}{\partial x_3}$$

$$= M_\infty^2 \left[(\gamma + 1) \frac{u}{U} + \frac{\gamma + 1}{2} \frac{u^2}{U^2} + \frac{\gamma - 1}{2} \frac{v^2 + w^2}{U^2} \right] \frac{\partial u}{\partial x_1}$$

$$+ M_\infty^2 \left[(\gamma - 1) \frac{u}{U} + \frac{\gamma + 1}{2} \frac{v^2}{U^2} + \frac{\gamma - 1}{2} \frac{w^2 + u^2}{U^2} \right] \frac{\partial v}{\partial x_2}$$

$$+ M_\infty^2 \left[(\gamma - 1) \frac{u}{U} + \frac{\gamma + 1}{2} \frac{w^2}{U^2} + \frac{\gamma - 1}{2} \frac{u^2 + v^2}{U^2} \right] \frac{\partial w}{\partial x_3}$$

$$+ M_\infty^2 \left[\frac{v}{U} \left(1 + \frac{u}{U} \right) \left(\frac{\partial u}{\partial x_2} + \frac{\partial v}{\partial x_1} \right) \right.$$

$$\left. + \frac{w}{U} \left(1 + \frac{u}{U} \right) \left(\frac{\partial u}{\partial x_3} + \frac{\partial w}{\partial x_1} \right) + \frac{vw}{U^2} \left(1 + \frac{u}{U} \right) \left(\frac{\partial w}{\partial x_2} + \frac{\partial v}{\partial x_3} \right) \right] \tag{4.5}$$

A first simplification can be done by dropping cubic terms on the right-hand side of the above equation. We have, then:

$$\left(1 - M_\infty^2\right) \frac{\partial u}{\partial x_1} + \frac{\partial v}{\partial x_2} + \frac{\partial w}{\partial x_3} = M_\infty^2 \left[(\gamma + 1) \frac{u}{U} \frac{\partial u}{\partial x_1} \right.$$

$$+ (\gamma - 1) \frac{u}{U} \left(\frac{\partial v}{\partial x_2} + \frac{\partial w}{\partial x_3} \right)$$

$$\left. + \frac{v}{U} \left(\frac{\partial u}{\partial x_2} + \frac{\partial v}{\partial x_1} \right) + \frac{w}{U} \left(\frac{\partial u}{\partial x_3} + \frac{\partial w}{\partial x_1} \right) \right].$$

$$(4.6)$$

The above equation still contains nonlinear quadratic terms. Further simplification is possible by dropping these terms, what corresponds to linearizing the potential equation of compressible flows. We obtain:

$$\left(1 - M_\infty^2\right) \frac{\partial u}{\partial x_1} + \frac{\partial v}{\partial x_2} + \frac{\partial w}{\partial x_3} = 0. \tag{4.7}$$

Linearization of Eq. 4.6 is based on the hypothesis posed by Eq. 4.1, which assumes that the perturbation components of the velocity, u, v, and w are small when compared to U. If we assume that U is of order $\mathcal{O}(1)$, then the perturbation components are much smaller, being, say, of order of a small parameter ε. So, are also the derivatives of the perturbation components. Since the right-hand side of Eq. 4.6 contains quadratic terms only, these terms are of order $\mathcal{O}(\varepsilon^2)$, whereas the left-hand side of that equation is of order $\mathcal{O}(\varepsilon)$. The right-hand side can thus be entirely dropped, and we end with the linear Eq. 4.7.

However, the transonic case, where $M_\infty \to 1$, namely, $1 - M_\infty = \varepsilon$, deserves special attention. In this case, the derivative $\partial u / \partial x_1$ can be of order $\mathcal{O}(1)$, while keeping all terms of the left-hand side of Eq. 4.7 of order $\mathcal{O}(\varepsilon)$. Indeed, in this case, the first term on the right-hand side of Eq. 4.6 becomes of order $\mathcal{O}(\varepsilon)$ and can no longer be neglected. We arrive this to a nonlinear equation valid for subsonic, transonic, or supersonic flows around slender bodies:

$$\left(1 - M_\infty^2\right) \frac{\partial u}{\partial x_1} + \frac{\partial v}{\partial x_2} + \frac{\partial w}{\partial x_3} = M_\infty^2 (\gamma + 1) \frac{u}{U} \frac{\partial u}{\partial x_1}. \tag{4.8}$$

Equations 4.7 and 4.8 can also be written in terms of the potential ϕ:

$$\left(1 - M_\infty^2\right) \frac{\partial^2 \phi}{\partial x_1^2} + \frac{\partial^2 \phi}{\partial x_2^2} + \frac{\partial^2 \phi}{\partial x_3^2} = 0 \tag{4.9}$$

$$\left(1 - M_\infty^2\right) \frac{\partial^2 \phi}{\partial x_1^2} + \frac{\partial^2 \phi}{\partial x_2^2} + \frac{\partial^2 \phi}{\partial x_3^2} = \frac{M_\infty^2 (\gamma + 1)}{U} \frac{\partial \phi}{\partial x_1} \frac{\partial^2 \phi}{\partial x_1^2} \tag{4.10}$$

Although Eq. 4.10 holds in the subsonic and supersonic regimes, the simpler one, Eq. 4.9, can be used in these regimes.

3 The Pressure Coefficient

We now obtain an expression for the Pressure Coefficient, defined by Eq. 2.16, in terms of the incoming flow velocity U and of the perturbation components u, v, and w. Rewriting Eq. 2.16 to include the three components of the velocity perturbation, we have:

$$
\begin{aligned}
C_p &= \frac{2}{\gamma M_\infty^2} \left\{ \left[1 + \frac{\gamma-1}{2} M_\infty^2 \left(1 - \frac{(U+u)^2 + v^2 + w^2}{2} \right) \right]^{\gamma/(\gamma-1)} - 1 \right\} \\
&= \frac{2}{\gamma M_\infty^2} \left\{ \left[1 - \frac{\gamma-1}{2} M_\infty^2 \left(\frac{2u}{U} + \frac{u^2 + v^2 + w^2}{2} \right) \right]^{\gamma/(\gamma-1)} - 1 \right\}.
\end{aligned}
$$

By developing the terms inside the square brackets with the binomial formula, and dropping cubic and higher order terms, what is consistent with the order of Eq. 4.6, we obtain:

$$
C_p = - \left[\frac{2u}{U} + \left(1 - M_\infty^2 \right) \frac{u^2}{U^2} + \frac{v^2 + w^2}{U^2} \right].
$$

In the case of two-dimensional planar flows, where perturbations are of first order, the quadratic terms in the above equation can also be neglected, leading to:

$$
C_p = - \frac{2u}{U}. \tag{4.11}
$$

4 Boundary Conditions

In the framework of inviscid flows, evolving in accordance with Euler's equation, the velocity at the interface with solids does not vanish, being parallel to the surface. Assuming that the solid surface is described by a function of the form:

$$
f(x_1, x_2, x_3) = 0,
$$

the boundary condition is expressed by:

$$\mathbf{v} \cdot \mathbf{grad}\ f \ = \ 0 \qquad \text{or} \qquad u_i \frac{\partial f}{\partial x_i} \ = \ 0.$$

Writing the full terms of the above equation, we have:

$$(U + u)\frac{\partial f}{\partial x_1} + v\frac{\partial f}{\partial x_2} + w\frac{\partial f}{\partial x_3} \ = \ 0.$$

Since $U + u \approx U$, we rewrite the above equation as:

$$U\frac{\partial f}{\partial x_1} + v\frac{\partial f}{\partial x_2} + w\frac{\partial f}{\partial x_3} \ = \ 0. \qquad (4.12)$$

In the two-dimensional case, Eq. 4.12 becomes:

$$\frac{v}{U} \ = \ -\frac{\partial f/\partial x_1}{\partial f/\partial x_2}.$$

At the solid surface:

$$df \ = \ 0 \quad \longrightarrow \quad \frac{\partial f}{\partial x_1}dx_1 + \frac{\partial f}{\partial x_1}dx_2 \quad \longrightarrow \quad -\frac{\partial f/\partial x_1}{\partial f/\partial x_2} \ = \ \frac{dx_2}{dx_1},$$

so:

$$\frac{v}{U} \ = \ \frac{dx_2}{dx_1},$$

where dx_2/dx_1 is the slope of the solid. The above equation states that the v perturbation component at the interface $v\,(x_1, x_2)$ is obtained by multiplying the slope dx_2/dx_1 by U:

$$v\,(x_1, x_2) \ = \ U\frac{dx_2}{dx_1}.$$

At this stage we make an additional approximation by assuming that:

$$v\,(x_1, 0) \ = \ v\,(x_1, x_2) \ = \ U\frac{dx_2}{dx_1}, \qquad (4.13)$$

an approximation valid for thin solids, aligned or almost aligned to the incoming flow.

5 Two-Dimensional Flow Past a Periodic-Shaped Wall

We consider now the case of a two-dimensional flow past a sinusoidal shaped wall in the form:

$$x_2 - \varepsilon \sin \alpha x_1 = 0, \tag{4.14}$$

where ε is the amplitude of the wall waves, $\alpha = 2\pi/\lambda$ and λ are the sinusoidal wavenumber and wavelength, respectively. Out of the transonic regime, the flow potential obeys the equation:

$$\left(1 - M_\infty^2\right) \frac{\partial \phi}{\partial x_1} + \frac{\partial \phi}{\partial x_2} = 0, \tag{4.15}$$

with boundary conditions:

$$\frac{\partial \phi}{\partial x_1}, \frac{\partial \phi}{\partial x_1} \quad \text{finite at infinity,}$$

and:

$$v\left(x_1, 0\right) = \left(\frac{\partial \phi}{\partial x_2}\right)_{x_2=0} = \left(U \frac{dx_2}{dx_1}\right)_{\text{wall}} = U\varepsilon\alpha \cos \alpha x_1. \tag{4.16}$$

5.1 Subsonic Flow Past a Periodic-Shaped Wall

We consider now the subsonic case, for which $1 - M_\infty^2 = m^2 > 0$. In this case, Eq. 4.9 becomes:

$$\frac{\partial^2 \phi}{\partial x_1^2} + \frac{1}{m^2} \frac{\partial^2 \phi}{\partial x_2^2} = 0. \tag{4.17}$$

This equation admits a solution with separation of variables, in the form:

$$\phi\left(x_1, x_2\right) = F\left(x_12\right) G\left(x_2\right).$$

Introducing the above solution in Eq. 4.17, we obtain:

$$\frac{1}{F}'' + \frac{1}{m^2 G} G'' = 0.$$

Since the first term of the above equation is a function of x_1 only, and the second one of x_2 only, the terms are constant, with:

$$\frac{1}{F}'' = -\alpha^2 \qquad \frac{1}{m^2 G}G'' = \alpha^2$$

the above equations admit solutions in the form:

$$F = A_1 \sin \alpha x_1 + A_2 \cos \alpha x_2$$

$$G = B_1 \exp(-m\alpha x_2) + B_2 \exp(m\alpha x_2).$$

Due to the requirement that the perturbations must vanish at $x_2 \longrightarrow \infty$, we impose $B_2 = 0$. At the wall, we have:

$$\left(\frac{\partial \phi}{\partial x_2}\right)_{x_2=0} = v(x_1, 0) = F(x_1)\left(\frac{\partial G}{\partial x_2}\right)_{x_2=0} = U \varepsilon \alpha \cos \alpha x_1. \qquad (4.18)$$

This condition is satisfied if $A_1 = 0$, and $-A_2 B_1 m\alpha = U \varepsilon \alpha$. The solutions for the potential, for the velocity components, and for the local pressure coefficient are thus:

$$\phi(x_1, x_2) = -\frac{U\varepsilon}{\sqrt{1 - M_\infty^2}} \exp\left(-\alpha x_2 \sqrt{1 - M_\infty^2}\right) \cos \alpha x_1 \qquad (4.19)$$

$$u = \frac{U\varepsilon}{\sqrt{1 - M_\infty^2}} \exp\left(-\alpha x_2 \sqrt{1 - M_\infty^2}\right) \sin \alpha x_1 \qquad (4.20)$$

$$v = U \varepsilon \alpha \exp\left(-\alpha x_2 \sqrt{1 - M_\infty^2}\right) \cos \alpha x_1 \qquad (4.21)$$

$$C_p = -2\frac{u}{U} = -\frac{2\varepsilon \alpha}{\sqrt{1 - M_\infty^2}} \exp\left(-\alpha x_2 \sqrt{1 - M_\infty^2}\right) \sin \alpha x_1. \qquad (4.22)$$

Several conclusions can be taken from the above solution of the flow:

1. We observe first that due to the exponential decay of the velocity to points far above and below the solid, the largest perturbation occurs at the solid interface. At the wall:

$$C_p = -\frac{2\varepsilon \alpha}{\sqrt{1 - M_\infty^2}} \sin \alpha x_1. \qquad (4.23)$$

2. The pressure coefficient increases with the Mach number proportional to $1/\sqrt{1 - M_\infty^2}$. In addition, the attenuation becomes weaker as the Mach number

increases. There is no drag force, since the pressure is in phase with the wall geometry, being thus symmetrical with respect to the crest of wavy wall.

3. The pressure coefficient can be expressed in nondimensional form:

$$\frac{C_p\sqrt{1 - M_\infty^2}}{\varepsilon\alpha} = -\exp\left(-\alpha x_2\sqrt{1 - M_\infty^2}\right)\sin\alpha x_1, \tag{4.24}$$

namely, the functional dependency of the pressure coefficient with ε, α, M_∞, x_1, and x_2 can be written in the form

$$f(\Pi_1, \Pi_2, \Pi_3) = 0,$$

where:

$$\Pi_1 = \frac{C_p\sqrt{1 - M_\infty^2}}{\varepsilon\alpha}, \qquad \Pi_2 = \alpha x_1, \qquad \Pi_3 = \alpha x_2\sqrt{1 - M_\infty^2}.$$

4. The meaning of small perturbations: from the expressions of the velocity perturbation components and the pressure coefficient it becomes clear that the hypothesis of flows with small perturbations, expressed by:

$$\frac{u}{U}, \frac{v}{U} \ll 1, \tag{4.25}$$

is equivalent to prescribing:

$$\frac{\varepsilon\alpha}{\sqrt{1 - M_\infty^2}} \ll 1. \tag{4.26}$$

By using the linearized Eq. 4.8, instead of Eq. 4.7 we assume that:

$$1 - M_\infty^2 \gg M_\infty^2(\gamma + 1)\frac{u}{U} = M_\infty^2(\gamma + 1)\frac{\varepsilon\alpha}{\sqrt{1 - M_\infty^2}}$$

or, equivalently, perturbations are small as far as:

$$\frac{M_\infty^2(\gamma + 1)\varepsilon\alpha}{\left(1 - M_\infty^2\right)^{3/2}} \ll 1.$$

5. We now derive the conditions for local sonic velocity, as maximum flow velocity: for $(1 - M_\infty) > 0$ but small, the derivative $\partial u/\partial x_1$ diverges and so does the right-hand side of Eq. 4.8. Since the derivatives $\partial v/\partial y$ and $\partial w/\partial x_2$ are small, these terms can be dropped and Eq. 4.8 becomes:

$$\left(1 - M_\infty^2\right) = M_\infty^2(\gamma + 1)\frac{u}{U}. \tag{4.27}$$

By replacing the term u/U by the expression obtained from Eq. 4.22, and noting that, at $M_\infty \longrightarrow 1$, $\exp\left(-\alpha x_2 \sqrt{1 - M_\infty^2}\right) \longrightarrow 1$, we have:

$$\frac{u}{U} = \frac{\varepsilon \alpha}{\sqrt{1 - M_\infty^2}} \sin \alpha x_1.$$

At points where the flow velocity is maximum and the local Mach number is equal to one we have $\sin \alpha x_1 = 1$. So, at these points:

$$\frac{u}{U} = \frac{\varepsilon \alpha}{\sqrt{1 - M_\infty^2}}.$$

Substituting the expression of u/U, given by the above equation in Eq. 4.27, we obtain the condition for maximum local velocity being sonic:

$$\frac{M_\infty^2 (\gamma + 1) \varepsilon \alpha}{\left(1 - M_\infty^2\right)^{3/2}} = 1.$$

6. To conclude, we investigate the conditions for validity of the boundary condition given by Eq. 4.13. From Eq. 4.21, we have the following expression for the vertical component of the perturbation velocity:

$$\frac{v}{U} = \varepsilon \alpha \exp\left(-\alpha x_2 \sqrt{1 - M_\infty^2}\right) \cos \alpha x_1.$$

Upon introducing the expression of x_2 in the above equation, we find an expression for the v component *at the boundary*:

$$\left(\frac{v}{U}\right)_{\text{boundary}} = \varepsilon \alpha \exp\left(-\varepsilon \alpha \sin \alpha x_1 \sqrt{1 - M_\infty^2}\right) \cos \alpha x_1$$

Development of the exponential in series leads to:

$$\left(\frac{v}{U}\right)_{\text{boundary}} = \varepsilon \alpha \cos \alpha x_1 \left(1 - \varepsilon \alpha \sin \alpha x_1 \sqrt{1 - M_\infty^2} + \ldots\right).$$

The vertical velocity at the boundary, given by the above equation, matches the boundary condition, as stated by Eq. 4.18 if:

$$\varepsilon \alpha \sqrt{1 - M_\infty^2} \ll 1.$$

In terms of maximum inclination θ of the wall, with $\tan\theta \approx \theta$:

$$\theta\sqrt{1 - M_\infty^2} \ll 1.$$

The above equation together with Eq. 4.26 define the limits of $\varepsilon\alpha$ or θ and of $\sqrt{1 - M_\infty^2}$ for validity of the theory of small perturbations in subsonic flow, presented in this chapter.

5.2 Supersonic Flow Past a Periodic-Shaped Wall

In the case of supersonic flows, $M_\infty^2 - 1 = m^2 > 1$, Eq. 4.15 becomes the wave equation, which is hyperbolic:

$$\frac{\partial\phi}{\partial x_1} - \frac{1}{m^2}\frac{\partial\phi}{\partial x_2}. \tag{4.28}$$

This equation admits a solution in the form of a sum of two arbitrary functions, $f(x_1 - mx_2)$ and $g(x_1 + mx_2)$:

$$\phi = f(x_1 - mx_2) + g(x_1 + mx_2). \tag{4.29}$$

For reasons that will become clear later, only the function f is needed, and g is taken as zero. The boundary condition is the same as in the subsonic case. At the wall:

$$v(x_1, x_2 = 0) = -\lambda\left[f'(x_1 - mx_2)\right]_{x_2=0} = -\lambda f'(x_1) = U\varepsilon\alpha\cos\alpha x_1,$$

where prime denotes derivative with respect to $\chi = x_1 - \lambda x_2$. In consequence:

$$f(x_1) = -\frac{U\varepsilon}{m}\sin\alpha x_1,$$

and we have, for the potential ϕ:

$$\phi(x_1, x_2) = f(x_1, x_2) = -\frac{U\varepsilon}{m}\sin\alpha(x_1 - mx_2)$$

or

$$\phi(x_1, x_2) = f(x_1 - mx_2) = -\frac{U\varepsilon}{\sqrt{M_\infty^2 - 1}}\sin\alpha\left(x_1 - x_2\sqrt{M_\infty^2 - 1}\right). \tag{4.30}$$

For the perturbation components of the velocity, and for the pressure coefficient, we have:

$$u = -\frac{U\varepsilon}{\sqrt{M_\infty^2 - 1}} \cos\alpha \left(x_1 - x_2\sqrt{M_\infty^2 - 1} \right) \tag{4.31}$$

$$v = U\varepsilon\alpha \cos\alpha \left(x_1 - x_2\sqrt{M_\infty^2 - 1} \right) \tag{4.32}$$

$$C_p = \frac{2\varepsilon\alpha}{\sqrt{M_\infty^2 - 1}} \cos\alpha \left(x_1 - x_2\sqrt{M_\infty^2 - 1} \right). \tag{4.33}$$

The main features of the supersonic flow over a wavy surface are:

1. The solution does not include an exponential attenuation, as found in the subsonic case. In consequence, the perturbations, also including pressure perturbations, do not decay with x_2. Instead, all perturbations keep the same value along the straight lines defined by:

$$x_1 = {}_2\sqrt{M_\infty^2} = \text{constant}. \tag{4.34}$$

These straight lines are denoted as *Mach lines* or *characteristics*, and are inclined at the Mach angle $\theta = \tan^{-1} 1/\sqrt{M_\infty^2}$, with respect to the incoming flow. The characteristics are independent of the particular boundary conditions, being due to the form of the solution for the supersonic potential, as given by Eq. 4.29.

   Mach lines associated with $\phi = f(x_1 - \lambda x_2)$ are inclined downstream the incoming flow, and originate at the wall, whereas those associated with $\phi = g(x_1 + \lambda x_2)$ are inclined upstream, and originate at infinite. Since the incoming flow is undisturbed these characteristic carry no perturbation and, consequently, $g = 0$ when the medium above the wall is unlimited.

2. The pressure coefficient on the wall is given by:

$$C_p = \frac{2\varepsilon\alpha}{\sqrt{M_\infty^2 - 1}} \cos\alpha x_1. = \frac{2}{\sqrt{M_\infty^2 - 1}} \frac{dx_2}{dx_1}. \tag{4.35}$$

Comparison with Eq. 4.23 shows that the pressure acting at the wall is shifted by a phase angle $\pi/2$. The pressure distribution is now anti-symmetrical with respect to the wall shape, leading to a drag. The average drag coefficient over a wall wavelength is given by:

$$C_D = \frac{1}{\lambda} \int_0^\lambda C_p \frac{dx_2}{dx_1} dx_1.$$

Inserting the expression of C_D, given by Eq. 4.35 on the above equation, we find:

$$C_D = \frac{2}{\sqrt{M_\infty^2 - 1}} \left(\frac{dx_2}{dx_1} \right)^2_{\text{avg}} a, \quad \text{where:} \quad \left(\frac{dx_2}{dx_1} \right)^2_{\text{avg}} = \frac{1}{\lambda} \int_0^\lambda \left(\frac{dx_2}{dx_1} \right)^2 dx_1.$$

Chapter 5
The Basic Equations of Compressible Fluid Flow

1 Introduction

This chapter summarizes the basic equations of compressible fluid flow. The complete time-dependent three-dimensional continuity and momentum equations are presented, including the constitutive equation of Newtonian fluids. The time-dependent Bernoulli equation for both compressible rotational and irrotational flows are discussed. The chapter presents the evolution equation of the circulation, where conditions for generation of vorticity and the necessary conditions for an irrotational flow to stay irrotational are discussed. One-dimensional mass, momentum, and energy equation are then presented.

2 Mass Conservation: Continuity

In order to derive the continuity equation in integral form we consider an immobile control volume V, simply connected, through which a fluid with specific mass ρ flows, with velocity $\mathbf{v}$. Let S be the external boundary of the volume and $\mathbf{n}$, the unitary vector, perpendicular to S and oriented outward, as shown in Fig. 5.1. The mass conservation principle states:

$$\begin{pmatrix} \text{Rate of accumulation of} \\ \text{mass in the volume } V \end{pmatrix} = - \begin{pmatrix} \text{Net balance of mass flow} \\ \text{outwards the volume } V \end{pmatrix}. \tag{5.1}$$

Application of the mass conservation principle, as stated above, leads to the integral form of the continuity equation in the form [2–6, 11, 13, 14]:

$$\int_V \frac{\partial \rho}{\partial t} \, dV = - \oint_S \rho \, \mathbf{v} \cdot \mathbf{n} \, dA. \tag{5.2}$$

© The Author(s), under exclusive licence to Springer Nature Switzerland AG 2019
J. Pontes et al., *An Introduction to Compressible Flows with Applications*,
SpringerBriefs in Mathematics, https://doi.org/10.1007/978-3-030-33253-2_5

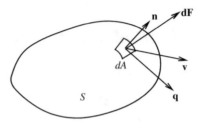

Fig. 5.1 A control volume to which the mass conservation principle is applied. **n** is the unitary vector, perpendicular to the surface S and **v**, the local velocity of the fluid element. The figure also shows the force applied to a surface element, and the vector heat flux crossing the element

The surface integral is transformed in a volume integral by employing Gauss' theorem. By doing so and applying Eq. 5.2 to an elementary volume we obtain the differential form of the continuity equation, which reads, in in Cartesian tensor notation, and vector form, respectively, as:

$$\frac{\partial \rho}{\partial t} + \frac{\partial \rho v_j}{\partial x_j} = 0,$$

$$\frac{\partial \rho}{\partial t} + \operatorname{div} \rho \mathbf{v} = 0.$$

3　Transport and Accumulation of a Scalar Variable

The integrals given by Eq. 5.2 represent, respectively, the rate of accumulation in a control volume, of the scalar variable mass, and the mass flow through the volume boundary S. The amount of the scalar variable mass in each element of the control volume is obtained by product of the local density—in the present case, the specific mass of the fluid, by the volume of the element.

The net flow of the scalar variable through the volume boundaries is obtained by integrating, along the entire boundary, the product of the local density of the scalar by the volumetric flow crossing the boundaries. The integrals given by Eq. 5.2 can thus be generalized to characterize not only the rate of accumulation, and the mass flow rate through the external boundary of the volume, but also, of any scalar variable, θ, having units of a scalar (mass, a component of the momentum, internal energy, kinetic energy, entropy, etc.). We represent the density of θ by c. If θ is the momentum component in a generic direction, associated with the fluid mass crossing the control volume boundary, c is defined by $c = \rho v_i$. If θ is the kinetic energy, $c = \rho v_i v_i / 2$. If the rate of accumulation of θ is equal to the net transfer of this variable to the control volume, the conservation of θ can be written in the form:

$$\int_V \frac{\partial c}{\partial t} dV = -\oint c\, v_j n_j dA.$$

By applying Gauss' theorem we obtain the conservation equation of the scalar variable θ in differential form:

$$\frac{\partial c}{\partial t} + \frac{\partial c\, v_j}{\partial x_j} = 0. \tag{5.3}$$

The conservation principle, as above stated, is employed to obtain the momentum equation of a compressible continuous medium.

4 Momentum Equation

The integral form of the Momentum equations is obtained by considering the transport of the scalar v_i, which is the velocity component along the direction of the generic unitary vector $\mathbf{e_i}$. In addition, momentum balance must take into account the forces applied to the volume, as shown in Fig. 5.1. We consider an immobile simply connected volume, embedded in the velocity field of a compressible continuous medium. The rate of change of the momentum contained in this volume comprises, in addition to the balance of the forces applied to the volume, the net flow of momentum through the boundaries of the volume [2, 3, 5, 6, 11, 13, 14]. Schematically (see Fig. 5.1):

$$\begin{pmatrix} \text{Rate of accumulation of} \\ \text{momentum in the control} \\ \text{volume } V \end{pmatrix} = - \begin{pmatrix} \text{Net flow of momentum} \\ \text{outward the volume } V \end{pmatrix}$$

$$+ \begin{pmatrix} \text{Balance of forces applied to} \\ \text{the control surface} \end{pmatrix}$$

$$+ \begin{pmatrix} \text{Balance of the body} \\ \text{forces applied to the ele-} \\ \text{mentary volumes of } V \end{pmatrix} \tag{5.4}$$

$$\int_V \frac{\partial}{\partial t}(\rho v_i)\, dV = - \oint_S \rho v_i v_j n_j\, dA + \oint_S \sigma_{ij} n_j\, dA + \int_V \rho g_i\, dV, \tag{5.5}$$

where σ_{ij} stands for the general component of the stresses tensor actuating on a surface element of the control volume. Surface integrals are transformed in volume integrals with Gauss' theorem, leading to the differential form of the momentum equation:

$$\frac{\partial}{\partial t}(\rho v_i) + \frac{\partial}{\partial x_j}(\rho v_i v_j) = \frac{\partial \sigma_{ij}}{\partial x_j} + \rho g_i.$$

Pressure is splitted from the stresses tensor in the form:

$$\sigma_{ij} = -p\delta_{ij} + \tau_{ij}.$$

Introduction of the above decomposition of the stresses tensor in the momentum equations results in:

$$\frac{\partial}{\partial t}(\rho v_i) + \frac{\partial}{\partial x_j}(\rho v_i v_j) = -\frac{\partial p \delta_{ij}}{\partial x_j} + \frac{\partial \tau_{ij}}{\partial x_j} + \rho g_i.$$

The left-hand side of the above equation can be simplified using the continuity equation. The result is:

$$\frac{\partial v_i}{\partial t} + v_j \frac{\partial \rho v_j}{\partial x_j} = -\frac{1}{\rho}\frac{\partial p}{\partial x_i} + \frac{1}{\rho}\frac{\partial \tau_{ij}}{\partial x_j} + g_i. \tag{5.6}$$

In vector form:

$$\frac{\partial \mathbf{v}}{\partial t} + \mathbf{v} \cdot \mathbf{grad}\, \mathbf{v} = -\frac{1}{\rho}\mathbf{grad}\, p + \frac{1}{\rho}\mathrm{div}\,\tau + \mathbf{g}. \tag{5.7}$$

In the case of Newtonian fluids like air, the constitutive equation of the viscous tensor is given by:

$$\tau_{ij} = \mu\left(\frac{\partial v_i}{\partial x_j} + \frac{\partial v_j}{\partial x_i}\right) + \lambda \delta_{ij}\frac{\partial v_k}{\partial x_k}. \tag{5.8}$$

In vector form:

$$\tau = \mu\left(\mathbf{grad}\,\mathbf{v} + \mathbf{grad}^T\mathbf{v}\right) + \lambda\,\mathrm{div}\,\mathbf{v}. \tag{5.9}$$

Euler's equation is obtained by dropping the viscous terms in the above momentum equations, which read in vector and in tensor notations:

$$\frac{\partial \mathbf{v}}{\partial t} + \mathbf{v} \cdot \mathbf{grad}\, \mathbf{v} = -\frac{1}{\rho}\mathbf{grad}\, p + +\mathbf{g}$$

$$\frac{\partial v_i}{\partial t} + v_j\frac{\partial \rho v_j}{\partial x_j} = -\frac{1}{\rho}\frac{\partial p}{\partial x_i} + g_i.$$

5   Bernoulli's Equation

Bernoulli's equation is derived from Euler's equation. As a first step we replace the convective term of the momentum equation, using the vector identity:

$$\mathbf{v} \cdot \mathbf{grad}\, \mathbf{v} = \mathbf{grad}\,\frac{\mathbf{v} \cdot \mathbf{v}}{2} - \mathbf{v} \times \mathbf{curl}\, \mathbf{v}.$$

The momentum equation becomes:

$$\frac{\partial \mathbf{v}}{\partial t} + \frac{1}{\rho}\,\mathbf{grad}\,p + \mathbf{grad}\,\frac{v^2}{2} - \mathbf{g} = \mathbf{v} \times \mathbf{curl}\,\mathbf{v}.$$

The term $\mathbf{g}$ and the pressure are incorporated to $\mathbf{grad}\,v^2/2$ resulting:

$$\frac{\partial v}{\partial t} + \mathbf{grad}\left(\int \frac{dp}{\rho} + \frac{v^2}{2} + gz\right) = \mathbf{v} \times \mathbf{curl}\,\mathbf{v}. \qquad (5.10)$$

The time-dependent Bernoulli's equation for irrotational flows takes the form:

$$\frac{\partial v}{\partial t} + \mathbf{grad}\left(\int \frac{dp}{\rho} + \frac{v^2}{2} + gz\right) = 0. \qquad (5.11)$$

According to Eq. 5.10, in time-independent flows, the vector:

$$\mathbf{grad}\left(\int \frac{dp}{\rho} + \frac{v^2}{2} + gz\right)$$

is orthogonal to $\mathbf{v} \times \mathbf{curl}\,\mathbf{v}$ (see Fig. 5.2), namely:

$$\int \frac{dp}{\rho} + \frac{v^2}{2} + gz = \text{Constant} \qquad (5.12)$$

along the surface to which the plane $\mathbf{curl}\,\mathbf{v}$ is tangent. If $\mathbf{rot}\,\mathbf{v} \neq 0$, Eq. 5.12 holds throughout the entire domain.

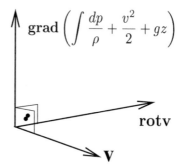

Fig. 5.2 Bernoulli's equation: when the velocity field is irrotational, the term $(\int dp/\rho + v^2/2 + gz)$ is constant along the surface having the tangent plane defined, at each point, by the vectors $\mathbf{v}$ and $\mathbf{curl}\,\mathbf{v}$. The angle between the two vectors may be different of $\pi/2$

6   Equation of the Circulation

This section deals with the mechanisms giving raise to the vorticity in fluids, namely, with the origins of the rotation of fluids. We consider an open, finite surface A of the space, bounded by a curve C, as shown in Fig. 5.3, and let $d\mathbf{l}$ be an element of this curve. The flow Γ of the vorticity vector across the surface A is defined as:

$$\Gamma = \int_A \mathbf{rot\,v} \cdot \mathbf{n}\, dA = \oint_C \mathbf{v} \cdot d\mathbf{l}.$$

This flow is equal to the circulation of the velocity vector along the boundary curve C, according to Stokes' theorem. The existence of a nonvanishing circulation is the signature of nonvanishing average curl in the domain A. We proceed in identifying the underlying mechanism affecting the evolution of the circulation around the boundary of a moving mass of fluid, namely we evaluate $D\Gamma/Dt$. We write:

$$\frac{D\Gamma}{Dt} \overset{\text{def}}{=} \oint_C \frac{D}{Dt}\mathbf{v} \cdot d\mathbf{l} = \oint_C \frac{D}{Dt} v_i\, dx_i = \oint_C \frac{Dv_i}{Dt} dx_i + \oint_C v_i d\frac{Dx_i}{Dt}$$

or

$$\frac{D\Gamma}{Dt} = \oint_C \frac{Dv_i}{Dt} dx_i + \oint_C v_i\, dv_i = \oint_C \frac{Dv_i}{Dt} dx_i + \oint_C \frac{1}{2} dv^2,$$

where $v^2 = \mathbf{v} \cdot \mathbf{v}$. The last integral copes with variations of a function along a closed curve. Since the initial and the final points of the curve coincide the function takes the same value in both points and this integral vanishes. We have, then:

$$\frac{D\Gamma}{Dt} = \oint_C \frac{Dv_i}{Dt} dx_i.$$

Taking into account that, according to the momentum equation:

$$\frac{Dv_i}{Dt} = -\frac{1}{\rho}\frac{\partial p}{\partial x_i} + \frac{1}{\rho}\frac{\partial \tau_{ij}}{\partial x_j}$$

Fig. 5.3 Circulation around a moving mass of fluid

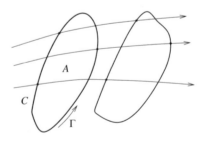

we can write:

$$\frac{D\Gamma}{Dt} = -\oint_C \frac{1}{\rho}\frac{\partial p}{\partial x_i}dx_i + \oint_C \frac{1}{\rho}\frac{\partial \tau_{ij}}{\partial x_j}dx_i$$

or either

$$\frac{D\Gamma}{Dt} = -\oint_C \frac{dp}{\rho} + \oint_C \frac{1}{\rho}\frac{\partial \tau_{ij}}{\partial x_j}dx_i. \tag{5.13}$$

The first integral on the right-hand side of the above equation vanishes in isoentropic processes. In general, $D\Gamma/Dt$ does not vanish when the entropy varies along the curve, due to either reversible processes like heating, or irreversible ones like mixing of masses with different temperatures, or with different chemical composition. The second integral in Eq. 5.13 copes with changes in the circulation due to viscous effects.

A corollary of the above result is *Kelvin's theorem*, stating that if thermodynamic irreversibilities and viscous effects are negligible:

$$\frac{D\Gamma}{Dt} = 0. \tag{5.14}$$

The last equation stresses the importance of irrotational flows, by showing that, in conditions where Kelvin's theorem apply, if **curl v** $\equiv 0$ in the domain, the flow stays irrotational as it evolves in time and in the space. An irrotational flow perturbed by a body immersed in the field keeps irrotational, no matter the geometry of the body is, as far as viscous effects are negligible.

7 Energy Equations

1. Kinetic Energy

$$\frac{D}{Dt}\left(\frac{v_i v_i}{2}\right) = -\frac{1}{\rho}\left(v_i\frac{\partial p}{\partial x_i} - v_i\frac{\partial \tau_{ij}}{\partial x_j}\right) + v_i g_i.$$

2. Total Energy Total energy is defined as the sum of internal plus kinetic. Balance of the total energy in a control volume as shown in Fig. 5.1 Schematically:

$$\left(\begin{array}{l}\text{Rate of accumulation of internal plus kinetic energy in}\\ \text{the control volume}\end{array}\right)$$

$$= -\left(\begin{array}{l}\text{Net rate of flow of internal}\\ \text{plus kinetic energy leaving}\\ \text{the control volume}\end{array}\right) + \left(\begin{array}{l}\text{Work per unit of time of the}\\ \text{forces applied to the surface}\\ \text{of the control volume}\end{array}\right)$$

$$+ \begin{pmatrix} \text{Work per unit of time of the} \\ \text{body forces applied to the} \\ \text{fluid in the control volume} \end{pmatrix} - \begin{pmatrix} \text{Net heat flow per unit of} \\ \text{time leaving the control vol-} \\ \text{ume} \end{pmatrix}$$

$$+ \begin{pmatrix} \text{Rate of heat generation inside the control vol-} \\ \text{ume} \end{pmatrix}.$$

Application of the energy conservation principle as stated above, to a control volume as shown in Fig. 5.1, leads to:

$$\int_V \frac{\partial}{\partial t} \rho \left(e + \frac{v^2}{2} \right) dV = - \int_V \frac{\partial}{\partial x_j} \rho \left(e + \frac{v^2}{2} \right) v_j \, dV$$

$$+ \int_V \frac{\partial v_i \sigma_{ij}}{\partial x_j} dV + \int_V \rho \, v_i g_i \, dV$$

$$- \int_V \frac{\partial q_j}{\partial x_j} dV + \int_V \dot{Q} \, dV,$$

where $\dot{Q}$ is the volumetric rate of generation of heat in the control volume. Upon transforming the surface integrals in volume integrals, and splitting the pressure from the stresses tensor we obtain:

$$\frac{\partial}{\partial t} \rho \left(e + \frac{v^2}{2} \right) + \frac{\partial}{\partial x_j} \rho \left(e + \frac{v^2}{2} \right) v_j = - \frac{\partial}{\partial x_j} p v_j + \frac{\partial}{\partial x_j} v_i \tau_{ij}$$

$$+ \rho \, v_i g_i - \frac{\partial q_j}{\partial x_j} + \dot{Q}. \qquad (5.15)$$

$$\frac{D}{Dt} \left(e + \frac{v_i v_i}{2} \right) = - \frac{1}{\rho} \frac{\partial}{\partial x_i} p v_i + \frac{1}{\rho} \frac{\partial}{\partial x_j} v_i \tau_{ij} + v_i g_i + \frac{\kappa}{\rho} \nabla^2 T + \frac{\dot{Q}}{\rho}.$$

3. Internal energy

$$\frac{De}{Dt} = - \frac{1}{\rho} p \frac{\partial v_i}{\partial x_i} + \frac{1}{\rho} \tau_{ij} \frac{\partial v_i}{\partial x_j} + \frac{\kappa}{\rho} \nabla^2 T + \frac{\dot{Q}}{\rho}.$$

4. Dissipation Function

$$\Phi = \tau_{ij} \frac{\partial v_i}{\partial x_j} = \frac{\mu}{2} \left(\frac{\partial v_i}{\partial x_j} + \frac{\partial v_j}{\partial x_i} \right)^2 + \lambda \delta_{ij} \frac{\partial v_i}{\partial x_j}.$$

5. Stagnation Enthalpy

$$\frac{Dh_0}{Dt} = \frac{1}{\rho} \frac{\partial p}{\partial t} + \frac{1}{\rho} \frac{\partial}{\partial x_j} v_i \tau_{ij} + v_i g_i + \frac{\kappa}{\rho} \nabla^2 T + \frac{\dot{Q}}{\rho}.$$

6. Entropy Equation

$$T \frac{Ds}{Dt} = \frac{1}{\rho} \tau_{ij} \frac{\partial v_i}{\partial x_j} + \frac{\kappa}{\rho} \nabla^2 T + \frac{\dot{Q}}{\rho}.$$

7. Internal Energy Equation

$$\frac{DT}{Dt} = -\frac{T}{\rho C_v} \left(\frac{\partial p}{\partial T} \right)_\rho \frac{\partial v_i}{\partial x_i} + \frac{1}{\rho C_v} \tau_{ij} \frac{\partial v_i}{\partial x_i} + \frac{\kappa}{\rho C_v} \nabla^2 T + \frac{\dot{Q}}{\rho C_v}.$$

8   The Equations of Steady One-Dimensional Flows

In this section we derive the equations applicable to steady one-dimensional flows, in channels with variable transversal section, where we assume that the velocity is uniform across any channel cross section. These equations are obtained from the integral form of the continuity, momentum, and total enthalpy, neglecting gravitational effects. The three-dimensional time-dependent integral equations read:

$$\int_V \frac{\partial \rho}{\partial t} dV = -\oint_S \rho v_j n_j dA \qquad (5.16)$$

$$\int_V \frac{\partial \rho v_i}{\partial t} dV = -\oint_S \rho v_i v_j n_j dA - \oint_S p n_i dA + \oint_S \tau_{ij} n_j dA \qquad (5.17)$$

$$\int_V \frac{\partial}{\partial t} \rho \left(e + \frac{v^2}{2} \right) dV = -\oint_S \rho \left(e + \frac{v^2}{2} \right) v_j n_j dA - \oint_S v_i p n_i dA$$
$$+ \oint_S v_i \tau_{ij} n_j dA - \oint_S q_j n_j dA. \qquad (5.18)$$

The first and the second terms on the right-hand side of Eq. 5.18 can be grouped. Since $e + p/\rho + v^2/2 = h_0$, we have for the two terms:

$$\oint_S \rho \left(e + \frac{v^2}{2} \right) v_j n_j dA + \oint_S v_i p n_i dA = \oint_S \rho \left(e + \frac{p}{\rho} + \frac{v^2}{2} \right) v_j n_j dA$$
$$= \oint_S \rho h_0 v_j n_j dA.$$

The total energy equation becomes:

$$\int_V \frac{\partial}{\partial t} \rho \left(e + \frac{v^2}{2} \right) dV = -\oint_S \rho h_0 v_j n_j dA$$
$$+ \oint_S v_i \tau_{ij} n_j \, dA - \oint_S q_j n_j \, dA. \qquad (5.19)$$

The integrals containing time derivatives vanish in steady state flows. We apply now the above equations to the steady flow through an elementary control volume, as shown in Fig. 5.1, and neglecting heat transfer along the flow direction. Let x be the coordinate along the flow direction, u the uniform velocity crossing the elementary control volume, and A the area of the duct transverse section.

8.1 Continuity Equation

Starting with the continuity equation, we have:

$$\rho u A + \frac{d}{dx} (\rho u A) - \rho A u = d\dot{m} = d(\rho A u) = 0, \qquad (5.20)$$

where $\dot{m}$ is the mass flow rate through the channel.

8.2 Momentum Equations

For the momentum equation, we write [9]:

$$pA + \left(p + \frac{dp}{2} \right) dA - (p + dp)(A + dA)$$
$$+ \rho A u^2 - \left[\rho A u^2 + d \left(\rho A u^2 \right) \right] + \tau_w dA_w \cos\theta = 0,$$

where τ_w is the shear stress applied by the duct wall on the fluid, and θ, the angle between the walls and the flow axis. We can simplify the term on the right-hand side of the above equation by noting that $\rho u A = \dot{m}$ is a constant and neglecting high order terms, to obtain:

$$A \, dp + d \left(\rho A u^2 \right) - \tau_w dA_w \cos\theta = 0.$$

This equation can be further simplified, taking into account that $\rho u A = \dot{m}$ is a constant, according to the continuity equation and that $dA_w \cos\theta = P \, dx$, where P is the perimeter of the duct transverse section.

$$A\,dp + \rho A u\,du - \tau_w P\,dx = 0.$$

Dividing the above equation by A, we have:

$$dp + \rho u\,du - \tau_w \frac{P}{A}dx = 0. \tag{5.21}$$

At this point, we define the Fanning factor by the relation:

$$f = \frac{-\tau_w}{\rho u^2/2}. \tag{5.22}$$

Using the above definition of the friction factor we rewrite Eq. 5.21 in the form:

$$dp + \rho u\,du + f\frac{\rho u^2}{2}\frac{P}{A}dx = 0. \tag{5.23}$$

Defining the hydraulic diameter by:

$$D_H = 4\frac{A}{P},$$

we rewrite Eq. 5.23 as:

$$dp + \rho u\,du + 4f\frac{\rho u^2}{2}\frac{dx}{D_H} = 0. \tag{5.24}$$

The flow pressure drop due to friction between the walls and the fluid is given by:

$$dp_f = 4f\frac{\rho u^2}{2}\frac{dx}{D_H}.$$

Another definition of the friction factor, denoted as the *Darcy's friction factor*, is usually used in the evaluation of pressure drop in pipes. Darcy's friction factor is defined through the relation:

$$dp_f = f_d\frac{\rho u^2}{2}\frac{dx}{D_H}, \tag{5.25}$$

from which, we have:

$$f_d = 4f.$$

Pressure losses in pipes are usually evaluated with Eq. 5.25, namely, with Darcy's friction factor, obtained from the well-known Moody's diagram. Throughout this book, we evaluate the pressure drop due to friction with the walls using the Fanning factor defined by Eq. 5.22.

The one-dimensional steady Euler equation neglecting gravitational effects is obtained from Eq. 5.24, by dropping the viscous term:

$$u\,du = -\frac{dp}{\rho}.$$ (5.26)

A case of particular importance occurs when the flow crosses a steady normal shock wave. The wave is sufficiently thin for neglecting friction forces between the flow and existing walls, and to assume that the area of the shock cross section is constant. Equation 5.21 simplifies and takes the form:

$$d\left(p + \rho u^2\right) = 0$$

or:

$$p_1 + \rho_1 u_1^2 = p_2 + \rho_2 u_2^2,$$ (5.27)

where subscripts 1 and 2 refer to the conditions just up and downstream the shock.

8.3   Energy Equations

In the case of a steady inviscid flow, without heat transfer between the duct walls and the fluid, the total energy equation becomes:

$$h_0 = h + \frac{u^2}{2} = \text{Constant}$$ (5.28)

or:

$$C_p T_0 = C_p T + \frac{u^2}{2} = \text{Constant } a,$$ (5.29)

where T_0 is the flow stagnation temperature. If heat is added per unit of mass of the fluid, at a rate dq, the stagnation temperature changes according to:

$$dq = C_p dT_0.$$

In the case of isothermal flows:

$$dq = u\,du.$$ (5.30)

8.4   *Entropy*

Entropy variations in a perfect gas can be evaluated by the following relation:

$$ds = \left(\frac{\partial s}{\partial T}\right)_p dT + \left(\frac{\partial s}{\partial p}\right)_T dp.$$

Considering the state equation of perfect gases and Maxwell's relation:

$$\left(\frac{\partial s}{\partial p}\right)_T = -\left(\frac{\partial v}{\partial T}\right)_p,$$

where v is the specific volume, we obtain:

$$ds = C_p \frac{dT}{T} - R\frac{dp}{p},$$

where R is the gas constant of air. Taking into account that $C_p - C_v = R$, and that $C_p/C_v = R$, we rewrite the above equation as:

$$\frac{ds}{C_p T} = \frac{dT}{T} - \frac{\gamma - 1}{\gamma}\frac{dp}{p}. \tag{5.31}$$

References

1. J.D. Anderson Jr., *Modern Compressible Flow with Historical Perspective* (McGraw-Hill, New York, 1982)
2. G.K. Batchelor, *An Introduction to Fluid Mechanics* (Cambridge University Press, Cambridge, 1994)
3. R.B. Bird, W.E. Stweart, E.N. Lightfoot, *Transport Phenomena* (Wiley, New York, 1960)
4. A.L. Coimbra, *Mecânica dos Meios Cointínuos* (Ao Livro Técnico, Rio de Janeiro, 1967)
5. L.D. Landau, E.M. Lifshitz, *Fluid Mechanics* (Pergamon, New York, 1959)
6. W.H. Liepman, A. Roshko, *Elements of Gas Dynamics* (Wiley, New York, 1957)
7. M. MacLean, Detailed derivation of Fanno flow relashionships (October 2013). http://wwimeracfd.com/professional/technote/fanno_flow.pdf
8. K. Masatsuka, *I do like CFD*, vol. 1, 2nd edn. (Lulu.com, 2013)
9. P.H. Oosthuizen, W.E. Carscallen, *Introduction to Compressible Fluid Flow*, 1st edn. (Taylor & Francis Group, Boca Raton, 2014)
10. J. Pontes, N. Mangiavacchi, *Fenômenos de Transferência com Aplicações às Ciências Físicas e à Engenharia – Volume 1: Fundamentos* (SBM – Sociedade Brasileira de Matemática, Rio de Janeiro, 2016)
11. P.J. Pritchard, J.W. Mitchell, *Fox and McDonald's Introduction to Fluid Mechanics*, 9th edn. (Wiley, New Jersey, 2011)
12. E. Rathakrishnan, *Applied Gas Dynamics*, 1st edn. (Wiley, Singapore, 2010)
13. H. Schlichting, K. Gersten, *Boundary Layer Theory* (Springer, Berlin, 1999)
14. F.M. White, *Fluid Mechanics* (McGraw-Hill, New Jersey, 2015)

© The Author(s), under exclusive licence to Springer Nature Switzerland AG 2019
J. Pontes et al., *An Introduction to Compressible Flows with Applications*,
SpringerBriefs in Mathematics, https://doi.org/10.1007/978-3-030-33253-2

Printed in the United States
By Bookmasters